THREE STEPS TO THE UNIVERSE

THREE STEPS TO THE UNIVERSE

from the SUN *to* BLACK HOLES *to the* MYSTERY *of* DARK MATTER

David Garfinkle & Richard Garfinkle

THE UNIVERSITY OF CHICAGO PRESS
Chicago & London

The University of Chicago Press, Chicago 60637
The University of Chicago Press, Ltd., London

Paperback edition 2010

Printed in the United States of America

19 18 17 16 15 14 13 12 11 10   2 3 4 5 6

ISBN-13: 978-0-226-28346-3 (cloth)
ISBN-13: 978-0-226-28348-7 (paper)
ISBN-10: 0-226-28346-1 (cloth)
ISBN-10: 0-226-28348-8 (paper)

Library of Congress Cataloging-in-Publication Data
Garfinkle, David.
Three steps to the universe: from the sun to black holes to the mystery of dark matter / David Garfinkle and Richard Garfinkle.
p. cm.
ISBN-13: 978-0-226-28346-3 (cloth: alk. paper)
ISBN-10: 0-226-28346-1 (cloth: alk. paper) 1. Sun. 2. Black holes (Astronomy). 3. Dark matter (Astronomy). I. Garfinkle, Richard. II. Title.
QB521.G29 2008
520—dc22

2008008659

CONTENTS

ACKNOWLEDGMENTS

We would like to thank the following people: our father, Norton Garfinkle, for suggesting our collaboration in the first place. Our wives, Kim Garfinkle and Alessandra Kelley, for putting up with us during the time we spent working on this book (actually for putting up with us at all). Alessandra for the illustrations. Werner Israel for a thorough reading of the book and vitally useful comments. The members of the Cosmology and Gravity program of the Canadian Institute for Advanced Research for useful information and insights on astrophysics and cosmology presented at their yearly meetings. And finally, Jennifer Howard for being the perfect editor (a startling combination of open-minded and merciless) and Erin DeWitt for her careful yet user-friendly copyediting. We don't deserve as much help as we got, and we are grateful for every bit of it.

Introduction

How do you know the Sun will rise tomorrow? What justifies our day-to-day confidence that ice is slippery, or that a match put to the gas on the burner of a stove will make fire appear?

We learn these things by experience. We have seen the Sun rise, slipped on the ice, and time and again set the gas alight. But how can we learn about those things that are not at hand—the Sun and stars, for example—and how can we learn about things that do not even give us light to see them by, such as the dark objects physicists talk about: black holes, dark matter, and dark energy? We will seek to find these things out by a journey through three worlds that exist in the minds of scientists: the world we see, the world we can find out about, and the world we think we know.

WHERE WE WALK

On our journey out into the cosmos, we will explore three different metaphoric universes: the perceived universe, the detected universe, and the theoretical universe. It is in the interaction of these three universes that scientific understanding is created. The perceived universe is what we experience every day. It is the world we see, hear, smell, touch, taste, and remember from what we have seen, heard, smelled, touched, and tasted. It is the world in which our minds live much of the time. In fact, it might seem that we live our lives in the perceived universe, but in reality we reach the boundaries of that world many times each day.

Suppose you call your friend using a cell phone. You see the phone keypad, feel the buttons as you press them, and finally hear your friend's voice. All are parts of the perceived universe, until you ask the question "How does it work?" Your friend is miles away and yet his voice is coming to you out of this little metal-and-plastic box that you hold to your ear. How is this happening?

An explanation requires knowing what the cell phone is doing and how it interacts with the world around you in order to let you chat away. Your friend's cell phone takes the vibrations in the air produced by your friend's voice and produces a radio wave with that pattern of vibrations in it. That radio wave is relayed to your cell phone, which uses it to produce vibrations in the air that you hear as your friend's voice. The sound you hear is not your friend's voice, but a replica constructed from radio waves by your cell phone. Note that this explanation uses a phenomenon, radio waves, that we cannot see, hear, smell, touch, or taste. How then do we know that radio waves exist? How could we demonstrate that what we have just said is really how cell phones work?

We have gadgets, machines such as cell phones and other radio receivers, that react to the presence of radio waves by producing effects we can perceive (the voice of your friend, the music on your favorite radio station). Using this kind of apparatus, we connect the unperceived aspects of the world (such as radio waves) with the perceived (such as sound). The things that we cannot sense directly, but that we determine to exist by indirect means (that is, by using an apparatus) are the detected universe. We live in this universe all the time just as we live in the perceived universe, though we usually pay the least attention to it of all three universes. We focus largely on the perceived, such as the voice heard through the cell phone, rather than the means by which the cell phone works.

This inattention to the detected universe is the largest gap between scientific thinking and nonscientific thinking. Even though we live in the detected universe, we tend to fold its effects into the perceived universe, which creates strange impressions and illusions. We look at our computer screens and act as if the Internet were really in front of us when, so far as most of us would define reality, it does not exist. The Internet is an aggregate of apparatus, hardware, and software that creates an illusion of existence. The apparatus of the Internet (millions of computers) communicates through phone and radio, relying on the detected universe for its existence, and creating perceptions for our perceived universes, such as web pages appearing on our screens.

This concealment of the detected universe is commonly called user-friendliness, the ability to use the unseen parts of existence without understanding how they work. User-friendliness is fine and good for everyday purposes, but to know what the universe is really like, to explore beyond our eyes and ears, it is necessary to jump the barrier of user-friendliness and discover the fascinating world beyond that layer of comfort. The universe that cannot be directly seen, like so many veiled things, has intrigues of its own.

Abandoning the comfort of things readily perceivable may not sound so pleasant, regardless of the promise of learning what lies behind. But the detected universe is more important in people's lives than the conveniences of cell phones and the Internet. It is where one of the greatest powers of the human mind resides, the power to discern what's really going on. Say a man falls down and hurts his arm. He would like to know if it is broken. He goes to the doctor, who takes an X-ray. The doctor looks at the X-ray picture and says the arm is indeed broken and that it will have to be put in a cast. The broken arm and the X-ray picture are things that can be seen and touched. But once the man wants to know "How does it work?" a voyage beyond the perceived universe must take place.

The X-ray machine produces a type of radiation called X-rays that are similar to visible light, but have such a short wavelength that our eyes cannot see them. Just as with light in an ordinary camera, when X-rays hit the film, they cause a chemical reaction that turns the film, once it is developed, into a different color from the parts of the film that have not been hit. The X-rays pass readily through skin and muscle, but not so readily through bone. The result of this is that in X-ray "light" bones cast shadows on the film, and when the film is developed, these shadows become the X-ray picture that our man and his doctor can review.

The above tells us not just how an X-ray machine works, but what it fails to do. X-ray pictures show us the difference between those things solid enough to block X-rays and those not so. If we accepted the crude statement "X-rays show you what's inside," we would not realize that an X-ray cannot easily distinguish between two kinds of materials that are transparent to X-rays and therefore cannot see many kinds of internal injuries. If you know how a thing works, you understand its limitations. And you may find yourself wanting to do more. The utility and limitations of X-ray photography prompted the invention of other means of examining the internal body, such as sonograms and MRIs.

These descriptions of cell phone conversations and medical X-ray diagnosis are mere sketches. A fuller explanation of cell phone conversations would give more details about how the components of the cell phone work, about how radio waves travel through the air, how the vocal cords and mouth produce sounds, how those sounds travel through the air, and how the ear perceives them. A fuller explanation of medical X-ray diagnosis would tell how the machine generates X-rays, why the film is sensitive to X-rays, and why bone tends to block X-rays more than muscle does.

Cell phones and X-ray machines come with owners' manuals that describe how to use them and even begin to describe their properties. Human beings come with no such manuals. Nor do trees, stars, hurricanes, or volcanoes. How then do we find explanations for how these things work? And once we know how they work, what can we do with them?

This, in a nutshell, is the endeavor of science, the attempt to comprehend and harness the world around us. The explanations are composed in part of things that we perceive or detect—but only in part. The rest is a set of mental constructs, called theories. This world of pure mental creation is the last of our three parts. The theoretical universe weaves the perceived and detected universes together to form a coherent image. Theories serve two different and seemingly opposed functions: first, as overarching explanations for how things work and why things happen; and second, as initial points of departure for scientific exploration and the creation of new ideas and knowledge.

The modern theory of electricity ties in to the fundamental structures of matter and energy and treats electricity as a flow of subatomic particles (electrons) that interact with the things they flow through. Detailed understanding of this theory has enabled scientists and engineers to create a wide variety of machines that manipulate flows of electrons, from electron microscopes to computers such as the one these words are being written on.

The theory of electricity and its application made such devices possible. But in experimenting with the theory and creating real-world objects using it, the theory itself was tested. If the experiments and devices had not worked as expected in the detected and the perceived universes, electrical theory would have been called into question. When a theory is questioned, it is put to experiment. Experiments are performed in the detected and perceived universes in order to put the theoretical universe to the test.

The process of science thus flows in a circle—through theory, detection, and perception. Theory guides detection and perception, perception queries detection, and detection challenges theory. This dynamic process, the meat and drink of science, is regrettably the part of science that is least commonly popularized. Theories are often talked about, observations and detections are sometimes discussed in conversations about science, but the real dynamic—the harmony of our three universes that makes science what it is—is largely hidden from the general public. This is not because scientists are a secret cabal who want to clothe their work in mystery. Rather it is because in many respects it is the hardest part of science to explain. We're going to try to do so because we think it's worth knowing. We think that the communication gap between scientists and the general public is unnecessary and damaging to those on both sides of this chasm.

We hope to bridge the gap by explaining the ways science is done, and not shying away from those parts of science that in the clichés of popular imagery have made nonscientists run screaming into the night. We also hope to draw the gap closer from the other side. There is a propensity in the sciences for a sense of superiority and something of an embrace of the image of the sage. Showing how scientists do what they do removes a veil of mystery, something that science is generally in favor of. In our more grandiose moments, we hope to do something to seal the gap between the general public and scientists. Bear in mind that this book is being written by a science fiction writer who has the habit of building universes and a professor of relativity who has the habit of taking them apart. Overreaching is a common vice we have (that and cheap humor, which there will also be a lot of in this book). In our saner moments, we'll settle for better communication across the gap, and better jokes.

But why try to bridge the gap at all? Why can't we—and scientists—simply dispense with the detectable and theoretical worlds and just live in the factual, perceptible world of what we can see, taste, touch, and so on? Well, let's have a look . . .

TWAIN, EINSTEIN, AND THE FACTS

> There is something fascinating about science. One gets such wholesale returns of conjecture out of such a trifling investment of fact. —MARK TWAIN

> I am enough of an artist to draw freely upon my imagination. Imagination is more important than knowledge. Knowledge is limited. Imagination encircles the world. —ALBERT EINSTEIN

Twain was of course an insightful humorist, able to crystallize uncomfortable thoughts into biting lines. His words suggest scientists should stick with the facts instead of spinning exotic theories and indulging in wild speculation. Einstein seems to be saying the opposite, that imagination, the spinning of fancies, is more important than knowledge of the facts. But the gap between these two is an illusion. The writer knew that facts are the roots of fancy; the scientist, that fancies reveal facts.

But let us take Twain's comment at face value, which is only fair to literature's greatest curmudgeon. Why can't we just stick to the facts in science? First, it is important to note that there are two kinds of facts in science: those that come directly from our senses (the perceived universe), and those for which experimental apparatus is needed (the detected universe). When a botanist counts peas in a pod, the number is part of the perceived universe. When a microbiologist measures the length of a bacterium using a microscope, that is part of the detected universe.

While at first we may have been blithe in our acceptance of this second universe, now we should be uncomfortable. How do we know what it is the microscope is revealing to us? How do we know that what we detect is as much fact as what we directly perceive? Don't we need some kind of theory about how the apparatus works? But then why should we believe that theory? How can we avoid the circular argument that says that detection confirms the rightness of theory and theory the rightness of detection? If the two universes beyond our senses are this much of a problem, why can't we get by with just the perceived universe? We can, as long as we only want to answer certain questions, like "Can you see how bright that thing up in the sky is?" but not questions like "Why is it so bright?" or even "What is that?" We *can* live without answering more than those questions. We can accept the limitations of our senses and simply exist in the perceived universe.

At least, we can try to not answer them. But human nature seems to be against us. For many people, the itch to know more than their senses can tell them is too great to ignore. This itch may be nothing more than nosiness. You may desire to know what your neighbors are doing behind their closed doors, or to have some idea of what is happening in a country halfway around the world. For this purpose, we have gossip (and its more organized cousin, news media). Gossip is not direct perception and decidedly not factual. But the itch to know is clearly prevalent enough to support all the newspapers, magazines, TV, and Internet websites that supply an endless round of stories.

The temptation to explore beyond what we know is so strong that some teachers of mental self-discipline focus a great deal on the art of not going to look (discussions of such practices belong in a different book, by different authors). But it is only by allowing one's curiosity free rein that knowledge can increase. If one explores the unknown in a disciplined way, drawing inferences in such a way as to ensure that they bear out the known data, it is science. This is another definition of science: disciplined nosiness. Scientists explore the unknown through careful use and testing of theory and by employing trustworthy tools. The detected universe is accessed with the tools of science. Using appropriate hardware (and software, since many of the tools are mental), we extend our reach beyond our hands, our step beyond our feet, and our detection beyond our senses. But in order to use them for such purposes, we have to know how our tools work and under what circumstances they can be relied upon.

Let's consider the microscope as one of the standard tools for going beyond our natural limitations. Here's the standard explanation of what microscopes are and why biologists use them: Owing to the optical limitations of the human eye, there are some objects so small that we cannot see them unaided. These objects include bacteria and viruses, which are responsible for many diseases. But more than that, all living things are made of cells, so comprehending the functions of cells is an essential component in understanding the workings of living things. Our basic understanding of living things relies on objects we can't see with our eyes alone, but the microscope solves that for us, giving us unseen facts. Well, maybe. Here's where scientists become careful.

For an object that is small but not too small to be seen, we can compare the view of the naked eye to the view through the microscope. The comparison shows that the microscope gives a magnified view of the small object. We then *assume* when looking at an object that can only be seen under the microscope that the microscope is only giving us a magnified view of that object. However, things are not quite this simple. The image in a microscope can be distorted by, for example, dust on the lens or ketchup on the slide. To understand the possible sources of distortion, we need to understand how the microscope works. We need to know our tools before we can use them properly. Here again is the loss of user-friendliness, in the need to understand the way a tool works.

The microscope is not terribly germane to the subjects we will be covering in later chapters, but much that is said of arrangements of lenses meant to show things that are small can be said about the telescope, an arrangement of lenses meant to see things far away. The telescope is a tool we will discuss a lot as we look out upon the wider universe. So here is the similar standard explanation for telescopes: There are many astronomical objects that are so far away that they cannot be seen with the naked eye. A telescope magnifies distant objects, gathers more light, and provides more resolution than our eyes are able to do.

As with the microscope, the image in a telescope can be distorted. These distortions can come from flaws in the manufacturing of the lens or mirror of the telescope. But they can also occur as a result of the nature of lenses and mirrors. Glass, after all, has its own optical and structural properties that might interfere with the image we are seeing. In astronomy, some of these distortions have their own names: chromatic aberration for the tendency of lenses to bend different colors of light by different amounts, and spherical aberration for the tendency of the simplest round mirrors to give a slightly distorted image. By knowing the sources of possible error and the characteristics of those sources, scientists can remove those errors that can be removed and correct for those that cannot.

Telescopes and microscopes show one aspect of the perceived universe's limits: the resolution of our eyesight. But the limitations of this sense that we rely upon more than the others are more extreme and varied than simple resolution. Visible light is a narrow band in the huge range of electromagnetic waves. All electromagnetic waves can be thought of as light, and they differ from ordinary visible light only in their wavelength. "Wavelength" is a term created by analogy; when looking at waves in the water, wavelength is the distance between successive places where the water is highest in a series of ocean waves. Light is a phenomenon of electricity and magnetism (more about this later), and in light, wavelength means the distance between successive high points of the electric field (again more about this later). In more practical terms, in light, wavelength is color; we perceive different wavelengths as different hues.

The electromagnetic spectrum, then, is simply a number line representing different possible wavelengths. The region of visible light looks to us like a succession of colors because we perceive different wavelengths of visible light as different hues. Other regions of the electromagnetic spectrum, the stretches we cannot see, have other names: radio waves, microwaves, and infrared for the light that has wavelengths too large for us to see, and ultraviolet, X-rays, and gamma rays for the light that has wavelengths too small. All these types of light have their uses, and each can be detected by appropriate apparatus. But if we limited ourselves to only the facts that we can perceive, we would never be able to see the wavelengths beyond the resolution of our eyesight.

We can only reach that vast band of invisible facts by creating devices to sense what we cannot and then to translate what we cannot see into what we can. It has been said that poetry is what is lost in translation. No doubt there are potential regions of visual art that we will never be able to create because we cannot look directly at the X-ray sky or at the infrared heat signatures of living things. But the loss to poetry need not be a loss to science. We can build X-ray telescopes and infrared-sensitive goggles. We can indirectly determine what's out there, by detection.

Returning to the question of facts, we might, understanding our own limitations and the benefits of tools for detection, accept the need for the detected universe, since there are facts we can only learn indirectly. We might still be tempted to stick as close to the facts as we can, however, without theorizing. This does not work for two reasons. One, various imperfections in a detector (like ketchup on a microscope slide or spherical aberration in a telescope mirror) can lead to distortions in the information that we receive. In the jargon of science, detectors give both "signal" and "noise." In order to detect accurately, we must be able to understand and correct for the sources of noise. This requires a theory of how the detector works. Two, even with the best apparatus or detector, we can only measure some things about the objects we want to understand. Between the limitations of our tools and the way the universe works, we are sorely restricted in what we can find out. This is especially true for objects that are very far away. Right now, for example, we are only just able to determine the existence of planets in other solar systems. We have no hope at present of detecting what these worlds looks like from ground level (assuming they have ground levels—most of them are gas giants).

What do we do about the things we can't measure? About the holes in the perceived and detected universes? We fill them with theories. We need theories about the whole system in order to fit the universes together. These theories should be as clear and coherent as possible, and in the service of clarity, they should be as simple as we can make them, even as they must simultaneously and consistently agree with the things we can perceive and detect.

The need for simplicity is a matter of sensible mental caution. It is easy to spin webs and stories that catch hold of the imagination and seem wondrous and glorious and capture scientists' hearts. Such theories because of their beauty can be hard to give up (the same way it can be hard to toss away any pretty thing). But a scientific theory must be something that can be abandoned by the very people who create and use it. It must be something that can be challenged, and if successfully challenged, it must fall. If people become too attached to a theory, they lose the purpose of science, which is to create theories that explain facts in a way that allows accurate prediction of future events and that enables scientists to construct tools that act as expected. Simplicity in making theories keeps the mind from too much attachment to grandeur—although, it must be said, some people become attached to simplicity, so even that guide must be used with caution.

To be useful, scientific theories should predict things we can measure. They are confirmed, or at least supported, when those measurements come out as predicted. Since progress in technology leads to better apparatus, hence more accurate testing, there are continual opportunities to confirm or negate theories. If an experiment comes out as the theory predicts, then we can have more confidence in the theory. If not—and if we can be confident that the experiment was correctly created and performed—then it is time to search for a better theory. The boundary between the detected universe and the theoretical universe is a moving line. Something that we can't detect today may become detectable tomorrow. Sometimes, as with radio waves, what cannot be seen the day before yesterday is seen yesterday and marketed today.

No matter how hard we may try to live in the world of fact alone, it cannot be done. We live in a three-tiered universe. The perceived universe, the detected universe, and the theoretical universe are built upon a foundation of what we can perceive with our senses. Each part of science takes three steps through these universes, moving from observation to detection to theory. Each step outward is an essential part of understanding the physical world; the step back makes the understanding part of our lives.

The usual textbook presentation of science emphasizes hypothesis and experiment with no distinction between experiments that can be done with direct perception and those that cannot. A hypothesis is just a temporary mental crutch that becomes a fact when confirmed by experiment or is discarded if experiment shows it to be false. We wish to show that the notions of detections and theories have more rich and robust roles to play in science than the textbook notion of hypothesis. In this book we are going to spend most of our pages on several topics in astronomy. While our focus is on astronomy, we will of necessity digress into other branches of science because the sciences are not really separate. The observations, detections, and theories of one branch have often illuminated the puzzles of another.

Some of the greatest discoveries in the history of science have been acts of unifying seemingly disparate phenomena. James Clerk Maxwell's creation of a theoretical connection between electricity, light, and magnetism is not only beautiful to contemplate but dramatic in its consequences. The Scottish physicist and mathematician Maxwell (1831–1879) created something fundamental to much of what we use. His greatest work, known as Maxwell's equations, can be written in four lines, but thanks to these four lines, we have dynamos and radios, televisions and lots of other gadgets. From the standpoint of the sciences, there are more selfishly practical implications to the interconnection. If you and I are both examining aspects of the same thing and you have developed a tool that makes your eyes better and I have a tool that makes my ears better, then, as they say on children's television, we can share!

Astronomy is a tricky science upon which to focus, but one that benefits greatly from the aforementioned sharing. In many of the sciences, we can directly manipulate the objects of study and can see how they respond. We can make chemical reactions in a test tube, dissect frogs, or drop weights and measure the time it takes them to fall. In astronomy we are confined, for the most part, to passive observation of distant objects. This is true despite the amazing successes of space exploration. The *Pioneer 10* spacecraft, launched in 1972, has traveled over 8 billion miles, yet it is still less than 1/3,000th of the distance to the nearest star. The Hubble Space Telescope observes the farthest objects in the known universe from its orbit a mere 375 miles above Earth's surface, its great clarity due to the fact that it is above the distorting effects of Earth's atmosphere. Except for our own solar system, astronomy involves looking, not going there. Astronomy relies on the theoretical universe to fill in the gaps between our sparse observations. Indeed, some astronomical objects are known only indirectly, their existence inferred from their effects.

This indirectness is perhaps most acute for two classes of astronomical objects: black holes and dark matter. Black holes are objects whose gravity is so strong that not even light can escape from them. Therefore we cannot see a black hole. Dark matter is matter that is not giving off light, but whose presence is inferred through its gravitational effects. This property, of being out there but not seen, gives black holes and dark matter their present air of mystery. These two along with the even harder to detect dark energy are the things we will examine in greatest depth and through them learn more about the indirect methods that make up science.

It would be a mistake to think that only such dark objects require indirect exploration. Many other astronomical objects that we don't consider mysterious are also really only understood in an indirect way. Consider the interior of the Sun. We see light coming from the surface of the Sun but can't see the inside. No space probe that we could make could survive the heat of the Sun to probe its center, but the interior of the Sun is where its power is generated. Therefore our understanding of a basic fact about the Sun, how it radiates energy, depends on its unknown inner regions. We must use theory and detection to learn about these.

In our first step out into the cosmos, we will look closely at what we know about the Sun. Along the way we will cover some controversies—one that threatened to destroy the theory of evolution in its infancy and another that puzzled astronomers and physicists for decades and was only solved once and for all in the last few years.

Once we have warmed up on the Sun (sorry about that), we will take a step further to black holes. Black holes are a natural consequence of Einstein's theory of relativity, and in a sense they are some of the simplest objects in nature. They are also natural consequences of other theories concerning the behavior of stars, and in particular the ultimate fate of massive stars. There are many indirect observations of black holes coming from observations of the stars and gas around them. These observations are about two kinds of black holes: those whose mass is only a few solar masses (one solar mass is the mass of the Sun) and supermassive black holes in the centers of galaxies with masses of anywhere from a million to a billion solar masses. Our third step will take us to a study of dark matter and dark energy, two phenomena even harder to detect than black holes. There we will demonstrate the existence of matter that cannot be seen but only inferred through its gravitational effects.

After that we will take a step back and draw not just our astronomical, but also our scientific understanding back down to Earth. We will step through several of the sciences in order to connect the three-tiered universe of knowledge to everyday life. What makes Mark Twain's line funny is the common idea that science is just the facts, ma'am, just the facts. But this is not how scientists themselves see and speak about what they are doing. Inside the tent, as it were, behind the circus of facts, the performers speak a very different language and dwell in very different universes from the readily perceptible. So come inside the three-ringed circus. But first, let's watch the sunrise.

Step 1

THE

SUN

Previous page: Sun storm. NASA/JPL 2007.

CHAPTER 1

Look. Don't Touch.

Each of us has our own perceived universe. It is a world personalized by our perceptions, by the acuity of our senses and the circumstances of our lives. A nearsighted person will have different perceptions from a farsighted one, for example. But the most vital difference between two perceived universes is the different locations of the perceivers.

Inside a perceived universe, we can distinguish two kinds of objects: distant things that can be seen, heard, or—rarely—smelled, and nearby things that can touch us and make us feel their presence. What category an object falls into depends on where we are and where the object is. A mountain from a distance is of the first kind, but a mountain beneath our feet is of the second. Similarly, a hurricane watched on television is nothing like a hurricane whipping the world up around you.

The sky is full of objects of the first category, distantly seen but not felt. But there is one great exception to this tactile dichotomy: the Sun. Far away as it is, the Sun has an intimate immediacy. It touches us, warms us, burns us. When it is day, our skin cooks in its presence; at night our bones chill in it its absence. Despite its dominance in our lives, despite this skin-close feel and eye-blinding light, we cannot reach out and touch it. We can observe the Sun, but we cannot pick it up and experiment on it. Yet its blatant effect makes the understanding of the Sun a scientific imperative. We cannot justly say we know the world around us if we do not know the Sun. Yet distance makes it difficult to discern how one is to find out the nature of the Sun.

In order to attempt to understand the Sun, ancient scholars worked with what they could observe. If we imagine ourselves in their perceived universes, seeing much but knowing little, we can conceive of how we might translate this observed presence into understanding. We would begin with what we perceive and what we care about. We see the light of the Sun, and we see the Sun moving across the sky. We feel its warmth, and we perceive that this warmth and the length of time each day that the Sun is present in the sky vary as the year turns.

We learn by experience that the Sun and seasons are bound together. Historically, this knowledge sat for most of humanity's time upon Earth, learned quickly but understood not at all. Lacking instruments of detection that could tell our ancestors about the how and why of these annual changes, people created elaborate theories that drew upon this paucity of facts. These theories were not inherently scientific, being largely theological and mythological in character and serving purposes beyond that of explanation. It is not, however, the purpose of this chapter to delve into the uses of solar mythologies. Rather we look at what comes from paring away the stories, endeavoring to detect aspects of the Sun, as well as to see how those detections result in the theories that summarize our present understanding of the Sun.

Three aspects of the Sun—light, heat, and motion—are the most obvious characteristics to try to detect. For centuries the last of these, motion, formed the fundamental component of astronomical theory and detection. Astronomers and others not scientifically inclined wanted to know how the objects in the sky moved, fascinated as they were by the progress and regress of the Sun and planets. Ancient sky watchers developed accurate records and theories for these motions. The Greeks and those who followed their theories were stuck on a piece of theory: the idea that all things in the heavens moved in perfect circles. They created quite complex maps of the solar system that had circular paths piled on top of circular paths in a desperate attempt to bolster their theories.

This view of celestial motion was changed radically in the sixteenth and seventeenth centuries by the Polish astronomer Nicholas Copernicus and the Italian astronomer and physicist Galileo Galilee. Copernicus came up with the notion that Earth rotates around its axis and revolves around the Sun and that the planets also revolve around the Sun. In hindsight, the Copernican model seems obvious: the most blatant fact about celestial motion is that all objects in the sky seem to go around Earth once a day. In the Copernican model, all these different motions have a single cause: the rotation of Earth. But this simple explanation comes with a price: as we write these words at the latitude of the northern United States, the rotation of Earth means that we are moving at a speed of about 770 miles per hour and don't feel a thing. To Copernicus's contemporaries, it seemed absurd to claim that we are moving at such a ridiculously high speed and don't even notice it. This difficulty was resolved using a concept that we usually associate with Einstein, but that Einstein rightly attributed to Galileo: the principle of relativity. This principle says that what we notice is relative motion, not absolute motion. That is, we notice differences or changes in speed, not speed itself. When the two of us take a drive at seventy miles per hour in a convertible, we see motion relative to the road and feel the wind due to motion relative to the air. But here when we are writing this, the desk, chair, and room are all moving at the same 770 miles per hour: with no relative motion, we don't notice anything.

The Copernican model was further refined by the observations of one astronomer and the theories of another. The first, Tycho Brahe—a Danish astronomer who lived from 1546 to 1601—was one of the greatest proponents of the detected universe. This did not mean that one would have liked to have known him. Tycho was given his own island by the king of Denmark to persuade him not to leave for Germany, and there Tycho established his observatory (including a printing press to produce and bind his manuscripts, thus making Tycho one of the pioneers of the vanity press). From all reports, he was not a pleasant human being, governing his domain like a dictator. Brahe spent decades painstakingly recording (and making his assistants painstakingly record) the positions of the Sun and planets. He accumulated data with a precision unknown to his predecessors.

One of Brahe's assistants, Johannes Kepler, combed through Brahe's data and realized that the motion of the planets was not circular. Kepler's work consisted of distilling simple, elegant theory out of painstaking, bulky, boring data. He took decades of detected work and from it created three laws of planetary motion.

The first law is that planets move around the Sun in elliptical orbits. This may not seem like a big change from the idea of circular orbits, after all an ellipse is just an elongated circle. But circles have a geometric elegance and an aesthetic that had captivated minds for two thousand years. After all, went the theory, the sky is Heaven, Heaven is perfect, circles are perfect, therefore the sky is made of circles. This is an early example of attachment to theory, the desire to not let go of an idea even if it is shown to no longer work. Kepler's challenge to this concept was an incredible theoretical innovation.

Kepler's second law is that planets do not move at the same speed throughout their orbits, but move faster the closer they are to the Sun. His third law calculates how long a year is on a planet based on its distance from the Sun.

Kepler's method is a two-step process: observation determining what can be detected, and then detection distilled leading to theory is one of the most important processes in all scientific endeavor. This kind of distillation still goes on today. Even though modern scientists have computers to help them, they still have to go over the data with a human mind to create and test theories. In some ways, the task can be even more difficult today: the detecting apparatus in such large-scale experiments as the Human Genome Project and the Sloan Digital Sky Survey are so efficient that they quickly generated a huge amount of data. This great bulk of information is now available for poor human minds to try and make sense of. Information overload is a phenomenon that was once only known to scientists, but now anyone who has done a search on the Internet and discovered that there are more than a million websites that might contain the information they want, or worse that the information is spread throughout some dozens of those million sites, can get an idea of what a headache data analysis and synthesis can be.

Isaac Newton later built on Kepler's laws (as well as the work of others) to create his laws of motion, which ushered in the ability to calculate and predict motions. More about Newton later. But his work allowed astronomy to become less motion intensive and brought about a shift in the nature and focus of astronomical data. Where astronomers in the past were concerned largely with motion, in more recent times (since the motion problems are largely solved), it is the light and heat of the Sun and other stars and the study of where that light and heat come from that formed the most significant parts of modern solar and stellar astronomy. Thus astronomy shifted focus away from concerns of motion toward concerns of energy and stellar evolution.

This kind of sea change is important to note if one wants to understand the history of the sciences. A new theory or a new piece of detection hardware or a new ability to create something in a laboratory may cause a shift in interest and attention. If something that was once difficult or impossible to detect (such as the internal structure of a cell) becomes easy (thanks to advances in microscopy), then a part of a scientific field may move from cutting edge into normal work and eventually into foundational work for later cutting-edge research often in a completely different direction. Continuing the example of cellular microscopy, the discovery of chromosomes and the DNA code (see the final chapter for a brief discussion of this) led to a radical change of interest in biochemistry that brought forth the Human Genome Project and all those biologists going over the information right now.

Back to the sky.

Ancient astronomers rarely asked where the Sun came from or where it would go. They saw it as eternal, or at least divine enough to only be killed by other divinities. We now know that the Sun has a life cycle: birth, childhood, maturity, old age, and death. This occurs on a time scale that dwarfs not just our lifetimes but that of all life on Earth. Nevertheless, there was a birth, there is a life, and there will be a death of the Sun.

The how and why of those events and of solar light and heat are now known to be results of one physical process: nuclear fusion, a heavenly process that only in the last few decades was brought down to Earth. For purposes of study, of learning the ways of the Sun, fusion reactors and H-bombs afford scientists the ability to create briefly the same interactions that occur in the hearts of stars.

First, a bunch of numbers derived from a large number of detections and calculations. For the moment, we'll just pull them out of thin air, or this silk top hat you can't see. The Sun is about 150,000,000 kilometers (93,000,000 miles) away from us, it has a mass of about 1,990,000,000,000,000,000,000,000,000,000 kilograms, its surface temperature is 5,800 kelvin (centigrade degrees above absolute zero), and the temperature in its center is 15,500,000 kelvin. It is putting out about 400,000,000,000,000, 000,000,000,000 watts of power. (To get a grip on watts of power, take a monthly electric bill, find the line that tells you how many kilowatt-hours were used, and divide this by the number of hours in a month: about 720, and then multiply by 1,000. Do this and you'll see that your usage is just a tiny fraction of the Sun's output.) These extremely large numbers can be overwhelming. The use of "astronomical" as an adjective meaning so large as to defy imagination comes about because of numbers like these. There are three antidotes to this astronomical alienation: put numbers in scientific notation, use specialized units, and ask the question "How do they know that?"

Scientific notation is the writing of numbers using powers of ten. We have $10^1 = 10$, $10^2 = 100$, $10^3 = 1{,}000$, and so on. Thus we can write the distance to the Sun as 1.5×10^8 kilometers, the mass of the Sun as 1.99×10^{30} kilograms, and the power output of the Sun as 4×10^{26} watts. The same numbers can be made to look smaller and more manageable by the use of this notation, although that really doesn't change how mind-boggling the scale is.

Specialized units are simply ways of measuring things like distance, mass, and time, chosen so that for the system under study they give manageable numbers. The units we use every day (miles, kilometers, pounds, kilograms, and so on) are fit for the human, not the stellar, scale. Astronomers define the astronomical unit (AU) to be the average distance between Earth and the Sun. This is a convenient unit for discussing the solar system. The distance between Earth and the Sun is by definition 1 AU. The planet Mars has an average distance of about 1.5 AU from the Sun, Jupiter has a distance of 5.2 AU, and even Pluto is only about 40 AU from the Sun (depending on where it is in its eccentric orbit). Similarly, astronomers define the solar mass to be the mass of the Sun and the solar luminosity to be the power output of the Sun. It turns out that the Sun is a typical star, and so solar masses and solar luminosities are convenient units for discussing the properties of stars. For example, the star Sirius A has a mass of 2.2 solar masses and a luminosity of 23.5 solar luminosities.

For reasons to be detailed later, a convenient unit for distance to stars is the parsec (about 3.1×10^{13} kilometers and, no, it wasn't made up for *Star Trek*). A convenient unit for sizes of galaxies is the kiloparsec (one thousand parsecs), and a convenient unit for distances between galaxies is the megaparsec (1 million parsecs). In these units, the distance to the nearest star, Proxima Centauri, is 1.3 parsecs, the distance to the center of our galaxy is 8 kiloparsecs, and the distance to the Andromeda galaxy is 0.77 megaparsecs.

The question "How do they know that?" and the answers to that question provide a much more complicated, but ultimately much more rewarding, antidote to the enormous size of astronomical numbers than the simple use of scientific notation or specialized units. This question is a major key in the understanding of science and will be posed frequently throughout this book.

How do astronomers come up with the numbers for the distances, masses, and luminosities of objects? Are those numbers the results of detections? If so, what apparatus is used and how does it work? Is there a theoretical component used in arriving at these numbers? If so, what is the theory and how well is it confirmed? We will answer these questions for all the numbers we dropped a few paragraphs ago.

We begin with the AU, the distance from Earth to the Sun. For reasons of geometry and optics, it is easier to measure relative distances than absolute ones in the solar system. Consider the planet Venus. Venus is closer to the Sun than Earth is, and consequently its position in the sky is never too far from that of the Sun. That is why Venus is known as the morning star and the evening star; it either sets not long after the Sun does or rises not long before the Sun does. Venus is said to be at "maximum elongation" when its position in the sky is farthest from that of the Sun. Now suppose that Venus is at maximum elongation and consider the triangle made by Earth, the Sun, and Venus (see fig. 1). The angle at Venus is a right angle, while the angle at Earth is something that we can measure by simple observation of Venus and the Sun. If we know two angles of a triangle, we know the ratios of the sides (trigonometry takes care of this). With a little multiplication, we find the distance from Venus to the Sun as measured in AU. A similar but more complicated method works for planets that are farther from the Sun than Earth is.

Our treatment makes two approximations: (1) We are treating the orbits of the planets as circles when Brahe's sweat and Kepler's theory show them to be ellipses. This oversimplification is present whenever we speak of "the" distance of the planet from the Sun as a single number rather than something that changes as the planet moves. (2) Though the orbit of each planet lies in a

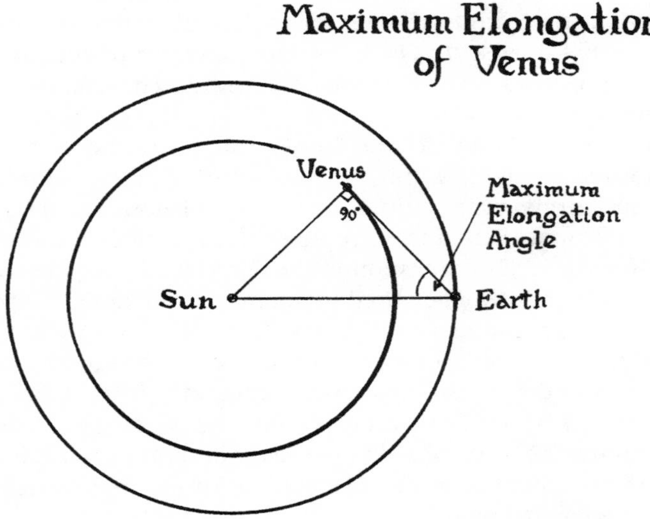

Figure 1

plane, in general the orbits of two planets don't lie in exactly the same plane.

We have two justifications for these simplifications. First, except for Mercury and Pluto, the orbits of the planets are pretty close to circles in the same plane, so unless we are doing very fine calculations, we are tolerably safe treating them as co-planar circles. And second, without this oversimplification the discussion would be much more complicated without producing much more useful figures. This last is important. Whenever scientists make approximations—and they make a lot of them—they always have to ask the question of whether or not those approximations make a significant difference in the answers they will get. When is a measurement significant? That depends. A difference of a few inches in a measurement of the height of a building rarely matters, but a difference of a few inches in the trajectory of a bullet can be the margin between life and death.

Scientists developed the concept of "significance" in measurement. Without belaboring the details, the significant part of a measurement is the part of it that is, at least somewhat, reliable. If you measure with a meter stick marked down to centimeters (1/100th of a meter), then the end of the measured object will generally fall somewhere between two markings. You can be sure of the measurement to two decimal places (that is, you are accurate down to 1/100th of a meter) simply by counting the number of markings, and you can make an estimate of the third decimal places (1/1,000th of a meter) by estimating the fraction of the distance between the two markings the end of the object is; but any attempt to guess the fourth or higher decimal place will really just be pure guesswork with no information. This sort of measurement is said to have three significant digits. Careful scientists always make sure that they present figures with the number of significant digits corresponding to the accuracy of the measurement made.

Using the methods of measurement and calculation given above, good approximations to the relative distances in the solar system were found. Armed with the relative distances, it was only necessary to measure one absolute distance in the solar system for all the absolute distances to be known. This may sound a little odd, but think about it like this: Using the triangulation method, it is possible to determine that Venus is 0.7 AU from the Sun, that Mars is 1.5 AU, and so on. Since an AU is the distance from Earth to the Sun, all that is necessary is to determine what an AU is and that gives us everything. If we can't find an AU directly, but we can find, say, the distance of Venus from the Sun directly, then we can determine what an AU is from that by simply dividing that distance by 0.7.

To find a single absolute distance, an idea called parallax was employed (see fig. 2). Parallax is an observed phenomenon, as well as a detected one. Try the following: Hold your left hand over your left eye and hold your right hand straight out with your thumb up so that it is lined up with some object on the far wall of the room. Now, using your left hand, uncover your left eye and cover your right eye. When you do this, the position of your thumb seems to shift. What is going on is that since your two eyes are at different positions, each eye views the thumb from a different angle. Half of the angle by which the thumb shifts is called its parallax. This angle is related to the distance between your eyes and the distance from your eyes to your thumb. Now, replace the two eyes by two points on Earth, the thumb by a planet, and the wall by the distant stars.

This sort of measurement was done in 1672 by two French astronomers, Jean Richer and Gian Domenico Cassini. Cassini is better known for his discovery of the gaps in Saturn's rings, and

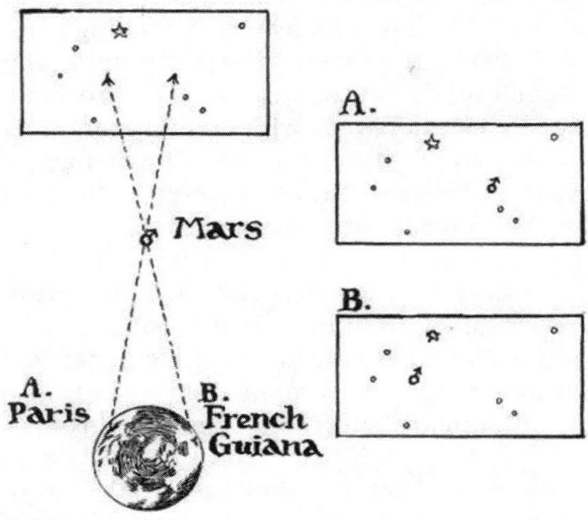

Figure 2a

the NASA Cassini mission to Saturn is named after him. He also did careful measurements of the position of Jupiter that allowed the Danish astronomer Ole Rømer in 1675 to do the first measurement of the speed of light.

Richer and Cassini each observed the planet Mars, one from Paris and the other from Cayenne. As viewed from these two vantage points, Mars seemed to have a slightly different position in the night sky as compared to the stars. Half of this angular difference was the parallax of Mars. While it is only a small angle, slightly less than 1/100th of a degree, using this parallax and the known distance between Paris and Cayenne (about 6,000 miles), Richer and Cassini were able to find the distance between Earth and Mars and thus the absolute distances within the solar system.

These days we have other ways of measuring distances in the solar system, for example, using radar or telemetry. Here's a description of one such method that requires a well-placed piece of detecting hardware: Whenever we send a signal to one of the Mars Rovers, we can measure the time between when we send the signal out and when we get a response back. Since radio

Stellar Parallax

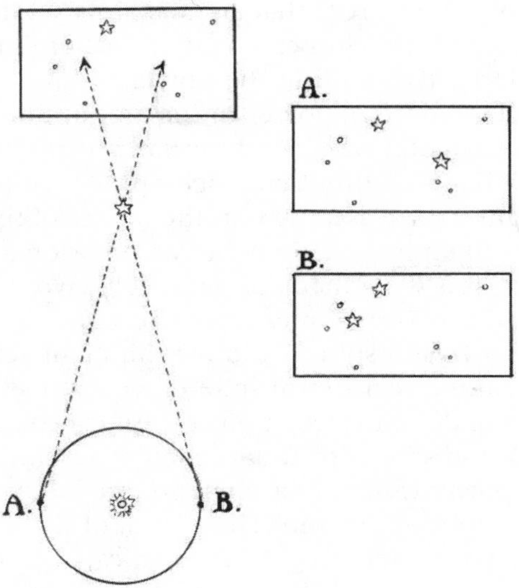

Figure 2b

waves are a kind of light, the signals travel at the speed of light (which has been measured to be about 186,000 miles per second or 300,000 kilometers per second), so we need only measure the time between signal and response, which gives us the time it takes a radio signal to make a round trip to Mars. Multiply half the time in seconds by 300,000, and we know how many kilometers it is to Mars. We need to be a little careful with this since the actual time consumed in the whole process is signal travel time + Mars Rover processing time for taking in the signal and creating its response + signal response travel time. If the middle term were large, it would mess up our calculations, but for a simple signal and a simple command, the time is in nanoseconds (billionths of a second)—insignificant on the time scale of minutes that we are dealing with. On the other hand, Galileo once tried to measure the speed of light using just this sort of method with an assistant (instead of the Mars Rover) and two lanterns (instead of the radio transmitters). He succeeded only in measuring the response time of the assistant.

Once the AU is known, from it follow several other numbers of "astronomical" size, for example, the size of the Sun. From observation it is easy to measure the angular size of the Sun. It is about half a degree. From that angular size and the distance to the Sun, the size of the Sun can be found (there's trigonometry again). Similarly, we can find the power output of the Sun by measuring the amount of power in the amount of sunlight that strikes a square meter of ground on Earth (one way to do this is to measure the amount of electricity produced in a photocell by this sunlight). If we assume that the power of the Sun comes out evenly in all directions, then we can envision a sphere 1 AU in radius that, if it were made of photocells, would capture the total power of the Sun. A physical sphere fitting that description is called a Dyson sphere and is a staple of science fiction, but even without a real Dyson sphere, we can use a theoretical Dyson sphere to do our calculation. We can calculate the surface area of the Dyson sphere in square meters and multiply that by the detected power caught in a single square meter to obtain the total power output of the Sun. These kind of theoretical experiments, where you imagine a giant impractical experiment (Dyson sphere of photocells) and then extract the essence of a useful experiment from it (one photocell and one act of multiplication), are one of the ways in which the detected and theoretical universes interact neatly, carrying us much further into knowledge than either alone.

There is one slight flaw in the above calculation: Earth's atmosphere absorbs some of the Sun's electromagnetic radiation, and the percentage of the radiation absorbed depends on the wavelength of the radiation. Thus, these experiments either need to be done in space, or the absorption of the atmosphere must be measured and corrected for. Even so, they can be and have been done in both ways, giving us the figures we pulled out earlier.

From the power and size of the Sun, we can find its surface temperature. Experiments in the laboratory with hot objects show that they give off electromagnetic radiation; the burner of an electric stove glowing red is an everyday form of this. The amount of radiation given off depends on the surface area and temperature of the object. Since we know the power and surface area of the Sun, we can calculate its surface temperature.

At this point we should become uncomfortable with this tenuous reasoning. The calculation of the surface temperature of the Sun involves the blithe combination of several different measurements. In addition, we are assuming that the properties of small hot objects in the laboratory can be extrapolated to an object, the Sun, that is much larger and far away. To a certain extent, this can't be helped. We can't put the Sun on a laboratory bench and put a thermometer on its surface.

However, we can find another indirect way to measure the Sun's temperature and see whether we get the same answer. This is a typical strategy in the detected universe. Any result needs to be checked using a different means than the original process in order to test whether the first result was correct. In the case of solar temperature, this other way also depends on the properties of hot objects and the radiation they give off.

Not only the power, but also the wavelength of the light emitted by a heated object depends on the temperature. Hotter objects give off light of a shorter wavelength. In light, the shorter the wavelength, the higher the energy; as we said before, wavelength of light manifests to our eyes as color. Heated iron becomes red-hot, but if we heat it even more, it becomes white-hot. By measuring the wavelength of the light coming from the Sun, we can determine its temperature. With two (indirect) ways to measure the temperature of the Sun, we can be more confident that each is correct. This sort of consistency check makes the indirect measurements less shaky than they would otherwise be. In the same way, the two different ways of measuring distances in the solar system, parallax and the speed of light, provide a consistency check. In fact, the first measurement of the speed of light was done indirectly using solar system distances and only later was it measured directly in the laboratory.

We can turn the above discussion around and look at the temperature of the Sun from the perspective of the theoretical rather than the detected universe. We begin with the theory that calculates how much light a hot object emits—a theory well tested in the laboratory, but we might worry that the theory will not hold for such a large and distant object as the Sun. What we then do is to make two different calculations, based on the two different sets of measurements discussed above, that—if the theory is correct—will each yield the temperature of the Sun. That the two calculations agree provides support for the theory, while the numbers calculated yield the temperature of the Sun. In this way, the detected and theoretical universes build off of and form checks upon each other. In the chapter on dark matter, we will see what science does when two such checks yield very different results.

Knowing the diameter of the Sun, we can determine its volume. We can then make a crude estimate of the mass of the Sun using mass = volume × density, with a guess for the density. Suppose we make a guess that the density of the Sun is about the same as the density of water. The answer that we get for the mass in this way is actually not far from the correct answer. But the problem is that we have no justification for the guess about density.

How do we do better? How do we "weigh" the Sun? In 1797 the English physicist Henry Cavendish came up with an indirect method. Cavendish initially set out to weigh Earth, but this also allowed him to weigh the Sun. Cavendish began with Newton's law of gravity, which states that given any two masses, each exerts a force on the other where the force is proportional to each mass and inversely proportional to the square of the distance between them. This law can be expressed in a compact way as the formula

$$F = GMm/r^2$$

Here F is the force, M and m are the masses of the two objects, r is the distance between them, and G is a constant known as Newton's gravitational constant. Newton came up with this formula as a simple and compact explanation for a host of phenomena from the fall of an apple to the motion of the Moon, from the tides to Kepler's laws of planetary motion. This formula allows us to compute the force that the masses exert on each other if we know the constant G. In order to use the formula to calculate forces, we need to know G. But how can G be found? This can also be made clear by the formula. Given two known masses a known distance apart, if we can measure the force between them, then the formula allows us to calculate G from that measurement. Once one such measurement is done, the value of G is known once and for all and the formula can then be used to find the force that any two masses exert on each other.

This sort of measurement is exactly what Cavendish did. His apparatus is shown in figure 3. Cavendish put two small lead balls on the ends of a rod and hung the rod from a thin steel wire. By measuring the amount the wire twisted when two large lead balls were brought close to the small ones, Cavendish was able to measure the gravitational force that the large balls exerted on the small ones. That is, in the formula for the law of gravitation, Cavendish measured all the quantities except G and then used

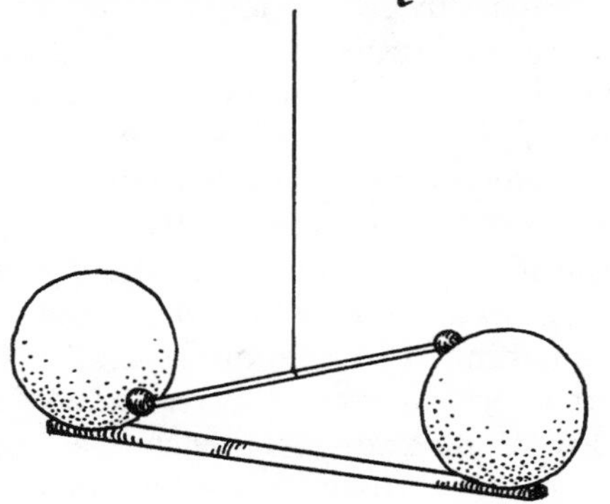

Figure 3

the formula to calculate G. What makes this measurement tricky is that the force of gravity is very weak. That may sound strange given how hard it is to defy gravity. But remember that with the formula above, one big mass (like that of Earth) can offset the inherent weakness of the force. Given two fourteen-pound bowling balls whose centers are three feet apart, the gravitational force that each exerts on the other is only about three-quarters of a billionth of a pound.

One consequence of Newton's law of gravity is another formula that allows us to find the mass of the Sun. This formula can be expressed as

$$M = rv^2/G$$

This formula refers to an object orbiting the mass M at a distance r and with a speed v. In words, the formula says that if we know the size and speed of the orbit (and if we know G), then we can compute the mass. In particular, we know that for Earth's orbit around the Sun, r is 1 AU. We also know that the speed is such that a circle of radius 1 AU is gone around in a time of one year. So from this information about Earth's orbit, we can compute the mass of the Sun (so this is the hat we pulled that number out of before). This little formula, which we will simply call the mass formula, is one of the most useful formulas in all of astronomy. Remember that in astronomy we do things by looking but not touching. Thus though we have no direct ways of finding masses, the mass formula provides us with a beautiful indirect way of measuring mass by simply looking at orbits. The beauty of this formula lies in its translation of easily detectable values (distance and speed) and from it producing hard-to-detect values (mass). This kind of sleight of equation is why science relies on mathematics as its court magician. This mass formula, now over three hundred years old, is not only the way that the mass of the Sun is found; it is also the way we find the masses of planets, stars, black holes, galaxies, and even dark matter.

The reasoning behind the Cavendish experiment sounds circular. To measure G, we assume the correctness of Newton's law of gravity. But if we don't know G, how can we test whether the law of gravity is correct? The idea is to test those predictions of Newton's law that don't depend on the value of G. In particular, look at the mass formula again. Since G is a constant and so is the mass of the Sun, and since the mass formula works for the orbit of any planet, one consequence of the mass formula is that rv^2 is the same for all planets. In other words, the farther a planet is from the Sun, the slower it is moving, and a planet has to be four times as far out to be moving at half the speed. This particular prediction was already known to be true: it was found by Kepler and was known as Kepler's third law. Thus Newton's law of gravity was checked by seeing that it provided an explanation for Kepler's third law (and as it turns out for the other two Kepler laws as well).

Now we know how we got all those numbers for distance, power, temperature, and mass of the Sun. Thank you, thank you. For our next trick . . .

CHAPTER 2

What's That Bright Thing Made Of?

Let us pop back to the perceived universe and look again at the Sun.

What do we see? A glowing, ball-shaped something that seems to move across the sky and is there roughly half the time. Many questions naturally arise in this simple observation, starting with the one we want to answer now: What is that thing? This question has been asked for much longer than human history. In some form or another, it probably predates our present human species and may even be older than anything we would care to call human. The previous statements are pure speculation, of course, not science. We cannot tell what our ancestors (human and otherwise) thought or asked, and we do have a tendency to project the questions that interest us onto others (especially if those others are not around to say "Who cares?"). In any case, the idea that they wanted to know what we want to know makes an interesting story. If we don't rely on it, we can use it as a story. But it is not science. For most of human history, the only answers to the question "What is that thing?" were stories. These stories were objects in the theoretical universe that had precious little connection to the universe of fact.

In these stories, the Sun has been described as a ball of fire, a hole in the sky, a god driving a chariot, a god running, a chariot being chased by a wolf, and on and on. These stories lacked connection to fact because there was no way to discern what the Sun truly was. The perceived and the theoretical universes had no connection, because until very recently, there was nothing about the Sun's composition within the *detected universe*.

As a side note, it is a common mistake to condemn those who made up stories about the Sun as being ignorant and foolish because they were not being scientific about it. But that is again a matter of projecting what interests us onto others. For the most part, they were not really trying to answer the question "What is that thing?"—rather they were using the question for other purposes. The Sun as a source of light, heat, life, as a marker of the day and year, served as a metaphor for understanding, for diligence, for the creation and destruction of things, for time, for power, for glory, for love. Not knowing what it physically was did not detract from its poetic value. And poets are not to be condemned for failing to put the disclaimer "I don't mean this literally" at the beginnings of all their poems.

Back to science: the reason it took so long to make any useful detections as regards the Sun's composition was a natural limitation on methods of observation and experiment. The Sun is a long way off, so we cannot simply grab a piece of it and stick it in a laboratory vessel. We have to use what comes out of the Sun to help us determine what it is. The only thing the Sun gives off that is simple to detect is light, lots of light. This, of course, makes the Sun easy to see, but not to analyze.

With the naked eye, one can only observe the large, coin-sized glowing disk in the sky. Once telescopes were invented, the Sun could be looked at more closely. Galileo himself was the first person to detect sunspots thanks to telescopes. Using these new wonder instruments, it became possible to make out the changing features of the Sun's surface. That it had features, changeable features, was an astonishing discovery, but it did little to answer the basic question of "What is that thing?"

Different tools coupled with subtler theories would eventually lead to our understanding the Sun's composition. But this would only come about through much diversion, digression, the melding of two branches of science, and the reshaping of the fundamental understanding of the nature of matter. The two most important tools were fire and prisms, and their usage in this instance came from the science of chemistry.

The question "What is this made of?" has been a staple of chemistry since it was alchemy. A chemist has a vast array of tests and a panoply of equipment that can be used to extract this information out of the sample. But those tests generally involve actually having the sample in front of you so that it can be subjected to these operations. And, as we said, we do not have the Sun to sample, only the light that comes from it.

Remarkably, there is a test for chemical composition that only involves observations of light. This test is called spectroscopy. Spectroscopy is partially familiar to anyone who has seen a rainbow. The light of the Sun is broken up by moisture in the air. White light is seen in the rainbow to be composed of light of many different colors, from red to violet. These different colors are in reality different wavelengths of light.

Light, as physicists figured out over much time and experimentation, consists of waves of changing electric and magnetic fields, analogous to the waves of water in the ocean. For water waves, we call the distance between successive waves the wavelength. A similar wavelength is defined for light, though there it is the distance between successive high points for the value of the electric field. Compared to ocean waves, light waves have an extremely high speed (3×10^8 meters per second) and extremely short wavelength (about 5×10^{-7} meters for yellow light).

In the visible part of the spectrum, red light has the longest wavelength, violet light the shortest. When light passes from one kind of transparent substance to another, the light gets bent. Light of different wavelengths are bent by different angles. In the case of a rainbow, sunlight goes through the air, enters a raindrop, bounces off the far side of the raindrop, and emerges into the air. Since light of different wavelengths emerges from the raindrops at different angles, we see the different colors of the rainbow in different parts of the sky. This produces the bands of the rainbow.

In the lab, the role of the raindrops is played by a prism or a diffraction grating. A prism is a triangular piece of glass or plastic. Light enters one side of the triangle and emerges from another bent at an angle, with the violet light bent at a slightly different angle from the red light. A diffraction grating is a thin piece of glass or plastic with many finely spaced parallel lines. Here the bending is accomplished by interference of the light that passes through (or reflects from) different parts of the glass or plastic. The angle of bending depends on the spacing between the lines and on the wavelength of the light.

Using these basic tools, spectrographs are built. A spectrograph is an instrument that consists of a prism or diffraction grating and a way of measuring the angle at which the light is bent. Thus a spectrograph takes in light, separates the light into its component wavelengths, and allows us to measure those wavelengths. The light that emerges from an object, spread out in the spectrograph, is called the spectrum of the object.

Anything that gives off light can have its spectrum measured. Naturally, the Sun was an object of such inquiry. What is observed in the spectrum of the Sun? Mostly a continuous rainbow from red to violet. But interspersed in the rainbow are dark "spectral lines" at particular wavelengths. These lines were first observed in 1802 by William Wollaston, a British scientist, best known as a pioneer of techniques of metallurgy (including the technique for obtaining pure platinum from platinum ore) and for the discovery of the elements palladium and rhodium. He did research in chemistry, mineralogy, crystallography, physics, astronomy, botany, physiology, and pathology.

The German chemist Joseph von Fraunhofer took Wollastan's work and expanded it; by 1814 he had discerned 475 of these lines (which are now called Fraunhofer lines) and had made an interesting discovery: one of these lines has the same wavelength as that of the yellow light produced when salt is sprinkled in a flame.

How do we get from a dash of salt in flame to what the Sun is made of? By turning up the heat. In the mid-nineteenth century, two Germans, physicist Gustav Kirchhoff and chemist Robert Bunsen, of Bunsen burner fame, discovered some very interesting properties of gases. A dense gas, when heated, gives off light of all wavelengths. However, a rarefied or thin gas behaves quite differently. When a rarefied gas that contains only a single chemical element is heated, it gives off only light of particular wavelengths. And every such sample of the same element will give off the same bands of light. This is called the spectrum of that element, and each element's spectrum is different from that of the other elements.

Furthermore, if light is passed through the cold rarefied gas of a chemical element, then only certain wavelengths of light will be absorbed by the gas, and these wavelengths are exactly the same ones that are emitted by the gas when it is hot. So if a light shines through a cold gas, the light that comes out on the other side will be missing the same colors that would have emerged if one were looking at a hot gas. Thus the spectrum of an element can be measured by using a spectrograph to look at the emission lines when the gas is hot or the absorption lines when the gas is cold. The spectrum of each element is unique to that element, so the spectrum can be regarded as a "fingerprint," a characteristic of that element distinguishing it from all the others. Fraunhofer's absorption line in the yellow part of the solar spectrum gave rise to the conclusion that the Sun contains sodium (one of the elements in salt) while measurements made by Bunsen and Kirchhoff showed that the Sun contains iron. The most prominent Fraunhofer lines correspond to spectral lines of hydrogen, iron, silicon, calcium, magnesium, and sodium.

The observation of the spectral fingerprints of elements exists in the detected universe. But before we can employ it, there must be a corresponding object in the theoretical universe that answers the question "Why do chemical elements behave this way?" That object of theory comes from the properties of light and atoms, which in turn require the broader theory of quantum mechanics.

To explain quantum mechanics, we begin with the concepts of counting and measuring. Counting is the simple process of assigning whole numbers to the idea of "how many" things of a certain kind there are. Given a pile of rocks, one can count how many rocks there are, picking them up one at a time and saying, "One rock, two rocks, three rocks . . ." until you run out of rocks. When you count, you get a whole number.

Measuring is a little more subtle. It involves things that are continuous instead of being separate, or discrete, as mathematicians call it. Continuous things include time, distance, area, volume, mass, and so on, things that could have any value, not just a whole number. To measure these things, we create units to represent one of that thing: for example, a meter is a unit of length. But we accept that things need not be whole units. We can have something that is 1.23 meters tall or an event that lasts 46.7 seconds. In our minds we distinguish between those things that can be counted and those that can be measured. In English we even use different phrases: "how many" for counted things, "how much" for measured things. How many cows? How much milk?

However, this distinction tends to blur when the things to be counted are very small and very numerous. Rice comes in individual grains; but we still talk about a cup of rice. Rice is really a counted quantity; but for practical purposes, we treat it as a measured quantity. Though we can perceive a single grain of rice, there are counted quantities so small that we cannot perceive them as counted and think of them as measured. However, detection or theory can tell us that something that we think of as a measured quantity is in fact a counted quantity. For example, chemistry tells us that water is really a counted quantity because there is a smallest quantity of water: a single H_2O molecule.

Quantum mechanics tells us that light is a counted quantity that comes in individual particles called photons. For light of wavelength *L*, each photon has the same energy *E* given by the formula

$$E = hc/L$$

Here, *c* is the speed of light and *h* is a number, called Planck's constant (around 6.626×10^{-34} joule seconds). Like the speed of light, Planck's constant is a fundamental constant of nature, a number that is an unchanging characteristic of the universe; it is used throughout quantum mechanics. Planck's constant is extremely small, so each photon of visible light has a very tiny energy. This is why we don't notice the individual photons from a lamp, any more than we notice the individual water molecules in a glass of water. The relationship between light seen as a wave (as we described it earlier) and light seen as a particle (as we are doing now) is one of those interesting aspects of the theoretical universe. There are characteristics of light that are most easily worked with when light is treated as a wave (wavelength, for example), and others (like the quantum characteristics we're talking about now) that are easier to work with when light is treated as a particle. The thing is that light is light; particles and waves are only models for what it is. Each model has aspects that work well under some circumstances. The question of whether it is really one or the other is a matter of people wanting to force things to be like their models. This is another example of people being too enamored of theory.

For our purposes, the most important feature of this formula is that light of a particular wavelength consists of photons of a particular energy. Thus quantum mechanics transforms the question we started with—"Why do chemical elements emit and absorb light of particular wavelengths?"—to the related question "Why do chemical elements emit and absorb photons of particular energies?"

To answer that question, we need to see what quantum mechanics says about the properties of atoms. Matter (except for the dark matter to be discussed in chapter 10) is made of atoms. From a chemist's point of view, there are only a few more than one hundred different kinds of atoms, but these atoms can combine in many different ways to form molecules. Molecules can be immensely complicated even if they are only made of a few simple atoms. Indeed, the complex chemicals that make life are mostly combinations of only five kinds of atoms: hydrogen, carbon, nitrogen, oxygen, and phosphorous.

This illustrates one of the aesthetically pleasing characteristics of our universe: it is harmonious, almost musical. At all levels of existence, from the smallest to the largest, a small number of things—like the famous eight little notes of a musical scale—can combine in a small variety of different ways to produce a large number of complex forms. Just as notes can combine into chords, into symphonies, so the small number of atoms above can combine in a small number of ways to produce molecules that combine to become life. We will look at this in more detail in our conclusion, but this principle of harmonious combination can be found throughout this book, as it will be found throughout the universe.

Atoms consist of a central nucleus composed of protons and neutrons, with a number of electrons orbiting around the nucleus. This model of the atom is often illustrated with a picture that looks like a solar system, with a nucleus in the place of the Sun and the electrons orbiting like planets. This image, however, has certain important limitations. In particular, just as quantum mechanics says that light waves come in particles called photons, it also says that electrons are described by waves and that in the presence of the nucleus only certain waves are allowed. (What we said before about the wave and particle properties of light also apply to electrons and every other particle; every particle can also be seen as a wave.) For our purposes, the crucial property of the allowed electron waves is that each one has a quantified (that is, counted) energy. Thus quantum mechanics tells us that atoms of a given chemical element can only have energies that belong to the list of possible energy states. In the solar system model of the atom, the energy of an electron is represented by how far out from the nucleus it is. (Remember this is only a model; distance is easier to visualize than energy, so we use distance as an image for energy, even though they are not the same.) The larger the energy, the "farther out" the electron.

The electron wave with the lowest energy is called the lowest energy level or the ground state, while the other allowed waves are called higher energy levels or excited states. As we said, intuitively one can think of the ground state as the orbit closest to the nucleus and the excited states as being orbits farther out; but this picture is only partially accurate. For purposes of careful science, it is better to forget about the notion of electron orbits and simply think in terms of energy levels, because just as "particle" and "wave" can distract from light, so "distance from the nucleus" can distract from energy. Models work only so far as they are useful, then you have to let them go.

If an electron in the ground state absorbs energy, it can jump to an excited state. If, on the other hand, an electron is already in an excited state, it can emit energy and fall back into its ground state. The energy is emitted as a photon, with the energy of the photon equal to the energy lost by the electron as it falls from excited to ground state.

More generally, if an atom is in an excited state, it can jump to any state of lower energy with the energy difference emitted as a photon. Similarly, an atom in one state can absorb a photon and change its state, but only if the energy of the photon corresponds to the energy difference between the two states. (If we had an electron in state 1, we could add a photon whose energy is the difference between the energies of states 1 and 2, 1 and 3, and so on, but not a photon whose energy was the half the difference between the energies of states 1 and 2, or 3.141592654 times the difference between the energies of states 1 and 3, et cetera. Only the photons whose energies add with the electron's energy to get the energy of one of the states of the atom can be absorbed by the atom.)

Now, let's go back a bit. The formula $E = hc/L$ tells us that the energy of a photon depends only on its wavelength, since h and c are both constants (numbers that do not change). But we already know that wavelength in light is the same thing as color. This tells us that for light, color and energy are essentially the same thing. Notice that using detection, we have made a strong connection between a very abstract idea (energy) and one of the most fundamental sensory components of the perceived universe (color). The quantum character of electron states in atoms tells us why atoms of a given type can only absorb light of certain colors and also why the wavelengths that can be absorbed are the same as the wavelengths that can be emitted. We now know what the fingerprints are, and we can safely use them for other purposes.

In our three-tiered model of the universe, we see that the theoretical universe of quantum mechanics offers an explanation for the detected-universe observation of chemical spectra. Without the theory, one would have to be careful in applying the tool of the spectra because one would not have cause to know under what circumstances one could use that tool. This did not stop the scientists of the mid-nineteenth century from using the tool without knowing why it worked. Truth to tell, the Sun's spectrum was determined decades before the theoretical foundation that justified the use of spectral analysis on the Sun. As we saw with Brahe and Kepler, sometimes detection precedes theory.

More broadly, one must be careful in using any tool to make sure it fits the usage (hammers, for example, make very poor ovens). Chemical spectra are only reliable when dealing with atoms that are floating free in gases, since complex molecules are capable of absorbing energy in other ways than just having their electrons excited (for example, the molecules can vibrate or spin). The only way to make sure that you have separate gases is to heat the materials enough to break down any bonds. Notice that this observation (heat breaking bonds) belongs again to chemistry's part of the theoretical universe from whence the astronomers borrowed it.

Fortunately, the Sun is very, very hot—far too hot to allow the chemical bonds of molecules to form. This means that most of the complicated questions of chemistry don't apply when trying to determine the composition of the Sun. This, incidentally, is itself an important observation. If the Sun is too hot for chemistry, we know already that all the things that depend on sophisticated chemical structures (life, for example) cannot exist on the Sun. This tells us that all we need to know in order to answer "What is that thing made of?" is which chemical elements exist in the Sun and in what abundances. In fact, most of the Sun is too hot even to allow the formation of atoms and instead consists of a gas of nuclei and electrons called a plasma. Nonetheless, we can still ask which nuclei exist and in what abundances.

Most of the light from the Sun comes from an outer layer called the photosphere, which is divided into the upper and lower photospheres. In the lower photosphere, the matter is sufficiently dense to give off light of all wavelengths. But the upper photosphere is cooler and less dense. Using a spectrograph to look at the Sun reveals absorption lines from the upper photosphere. These lines have certain wavelengths that can be compared to those found from chemical elements in the lab to see what the Sun is made of. So at last we can have the answer to "What is that made of?" which is a significant part of "What is that thing?"

This process found that set of gases in the Sun mentioned above. It also produced something else that caused problems for some time. Solar spectroscopy shows a spectral line that did not correspond to any chemical element that was known at the time these experiments were first done. This line was first observed in 1868 by the French astronomer Pierre Janssen as an emission line seen during an eclipse. Janssen showed singular devotion to studying eclipses. In 1870 when Paris was besieged during the Franco-Prussian War, Janssen escaped from the city in a balloon to observe an eclipse in Africa (but his view of the eclipse was blocked by clouds).

The same spectral line was observed also in 1868 by Joseph Lockyer, who hypothesized that it belonged to a hitherto-undiscovered chemical element. Lockyer called this element helium after the Greek word for the Sun, Helios. It may seem surprising now, when helium-filled balloons are at every child's birthday party, but at the time helium was a mystery in much the same way that dark matter is today. Here was a substance that had never been seen on Earth, was only seen in the Sun, and was only known indirectly through the presence of its spectral line. Given the evidence, one could believe in the existence of this mysterious "sun stuff," or instead could conjecture that there was some other explanation for the mysterious yellow line.

This question lay unanswered for nearly thirty years, sitting as a break in the theoretical universe. Was there some unknown element helium? Or was there something about using a laboratory technique (spectroscopy) on a distant object (the Sun) through the expanse of space that created some peculiar artifact? Without some separate means of finding out—that is, some detection that was not solar spectroscopy—the theoretical question could not be resolved.

In 1895 this issue was settled by the British chemist Sir William Ramsay, who had been a student of Bunsen's. Doing regular chemical analysis, Ramsay found the same spectral line, the line of helium, in a gas produced by heating the mineral cleveite. This independent, non-astronomical detection of the same element confirmed the existence of helium and settled the argument. (Not one to rest on his laurels, Ramsay went on to find all the other chemically inert gases—that is, those that don't tend to combine with other elements: neon, argon, krypton, xenon, and radon.) This not only confirmed helium's existence, but also helped to justify the use of spectroscopy in astronomy and made it possible to use it with greater confidence on more distant objects. In the early twentieth century, it was found that helium is often contained in certain deposits of natural gas. This is still the most abundant source of helium and has given us safe balloons and airships.

Using spectroscopy, astronomers have determined that most of the Sun is made of hydrogen and helium, with hydrogen accounting for 74 percent of the mass of the Sun, helium 25 percent, and all the other elements put together only 1 percent. (The reason for this has to do with big bang cosmology and will be explored in the chapter on dark matter.) So if helium is so common, why isn't it more prevalent on Earth?

Two of helium's properties account for this dearth of the gas: lightness and inertness. The atoms or molecules of a gas are always in motion. Earth, like all gravitating bodies, has what is called an escape velocity, a speed at which an object if it is moving away from the body can escape from Earth's gravitational pull and fly off into space. In the atmosphere, those gas molecules moving faster than the escape velocity get away, and thus each gas in the atmosphere is slowly leaking into space. How slowly? This depends on the average kinetic energy (energy of motion) of the atoms or molecules, and on their mass. We have a more everyday word for the average kinetic energy of atoms and molecules; we call it temperature. The hotter something is, the faster its atoms and molecules are moving; the colder something is, the slower these move.

Smaller atoms and molecules will be moving faster than larger ones even if they have the same temperature. Kinetic energy depends on both mass and speed ($mv^2/2$, if you must know, where m is the mass of the object and v is its speed. Nosy aren't you?), so to have the same kinetic energy, the less massive atom must be moving faster. If an atom in the atmosphere is light enough and the temperature high enough, it may be moving fast enough to escape Earth's gravity. On average, therefore, more light atoms will escape than heavy ones. Helium is the second lightest element (hydrogen is the lightest) and has a large enough average speed at any earthly temperature that our atmosphere cannot retain it. (Note that this does not mean that the average speed of a helium atom is above the escape velocity, simply that even that small percentage of helium atoms with speeds above the escape velocity is enough that any helium in the atmosphere gradually leaks away.) The planets Jupiter and Saturn are more massive and colder than Earth. Their gravity is strong enough to retain helium, which is a significant component of these planets.

Since hydrogen is even lighter than helium (one atom of hydrogen is about one-quarter the mass of that of an atom of helium), one might wonder why Earth contains so much more hydrogen than helium. This has to do with their chemical properties. Unlike hydrogen, helium does not combine with any other substance and therefore helium is only found as free-floating helium atoms. In contrast, hydrogen bonds with many other atoms, most famously with oxygen to form water. Earth's atmosphere does not retain hydrogen gas, which is lighter than helium; but it does retain the much heavier water. It also retains all the heavier hydrogen-bearing objects: people, for example. Hydrogen stays because it is stuck to other things; helium leaves because it is not.

Helium is quite useful, not only for balloons, but also because it never becomes solid, even at the lowest temperatures. It is used as a coolant in the refrigerators that achieve ultra-low temperatures very near absolute zero. As the second lightest element, it also plays a key role in both atomic physics and chemistry. However, for our purposes, the striking thing about helium is that though it makes up fully one-quarter of all the (ordinary non-dark) matter of the universe, it was undiscovered and completely unknown for almost all of human history. In this we see parallels to the mystery of dark matter. (In literature this use of something eerily similar to something that will be seen later is called foreshadowing. In science, it just happens that way sometimes.)

CHAPTER 3

Fusion: Energy Hidden in Plain Sight

In the perceived universe, one of the Sun's most obvious characteristics is its warmth, along with the observation that in general the day is warmer than the night. The one-line theory "The Sun heats Earth" may even be older than recorded history. Repeated experimentation, each and every day, confirms this theory. But why is the Sun hot? Theories that attempted to answer this question began with basic experience and extended it. The Sun looks like a fire, and fire needs something to burn in order to generate its heat. So what is the Sun burning?

Curiously, some of the earliest ideas actually ran contrary to this question. Many philosophers thought the Sun was an eternal source of heat that needed no fuel and would never burn out. In these theories, the constituents of the Sun were considered to be unlike those of Earth, the Sun being made of eternal and at least semi-divine materials. Through most of human history, it was natural to assume that the properties of the heavens were quite different from those of Earth. This assumption was rocked when Galileo using his telescope found mountains on the Moon and when

Newton found that the same law of gravity explains both the fall of an apple and the orbit of the Moon. These discoveries led to the opposite view: the uniformity of nature, which is the idea that the universe works the same everywhere, and that while the configuration of the universe can change over time and across space, the explanations apply everywhere and at all times. The uniformity of nature is a fundamental theory that underlies all modern scientific theories. In science, if there is an exception to something, there has to be a uniform theory that encompasses the thing and the exception.

Looked at from the view of this theory, the heat of the Sun must be caused by some form of fuel being consumed, since that's the only way we know of for things to burn. It is perfectly possible to live without accepting the uniformity of nature, but it is not really possible to predict anything, because if the way the world works changes in an unpredictable fashion, then events will arise unpredictably. This is not enough reason to accept the uniformity of nature. However, any theory made using the assumption of the uniformity of nature can be put to experimental tests. That these theories pass such tests gives us confidence that nature really is uniform.

The matter of the Sun's fuel connects two questions into one: "What is the source of the Sun's power?" and "How old is the Sun?" These questions are related because however the Sun is generating power, it must be using up fuel. However old the Sun is, it cannot be older than the time it would take to use up all that fuel. Since the properties of the Sun are so large as to be called astronomical, we might expect a large age too. However, recall that two of these large numbers are the mass of the Sun and the rate at which it is consuming fuel. The time it takes the fuel to be used up is the ratio of these two astronomical numbers, and the ratio of two very large numbers may not be large. After all, 2 million divided by 1 million is only 2.

When confronted with a problem of what is doing something, the first thing one should do is check if there is something one already knows of that might be responsible. So what are the known candidates for solar fuel sources? When we heat our homes, many of us burn natural gas. In a furnace supplying 20,000 watts of power, a kilogram of natural gas is consumed in about forty-five minutes. If the Sun got the same amount of energy out of a kilogram of its fuel as a furnace gets out of a kilogram of natural gas, then with a mass of 2.0×10^{30} kilograms and generating power of 3.8×10^{26} watts (to warm the entire solar system), from these numbers it can be computed that the Sun would burn up all its fuel in about ten thousand years. Similar numbers would result if instead of natural gas we were to use gasoline, alcohol, coal, wood, or jelly donuts (in this case "burned" inside the body to supply us with warmth and power).

This time scale was comfortably long when scientists' notion of the age of Earth was influenced by the biblical time scale of a scant few thousand years for the age of the universe. However, this calculation created a crisis in the nineteenth century with the rise of geology and the development of the theory of evolution. Geological change is very slow. Wind and water can wear down mountains, but it takes a very long time (millions of years). A river can dig the Grand Canyon, but this, too, takes a long time. In the nineteenth century, geologists began to estimate these time scales and arrived at a notion of the great age of Earth. Evolutionary change is also very slow. Evolutionary change (as we will discuss in later chapters) happens in incremental steps generated randomly and then tested by selection. Using statistics and paleontology, it is possible to make decent estimates of the time necessary for large-scale evolutionary changes, and these as well point to a time scale in the millions of years to get to where we are now.

There was a conflict of theories from different sciences, a war in the theoretical universe (it was a war fought with pen scratchings, but the subject matter was dramatic). The theory of evolution and the discoveries of geologists require that Earth have been around for at a minimum many millions of years. But if Earth needs the Sun for heat, then the Sun had to be fueled by something that would produce the measured energy while consuming much less fuel per unit amount of energy delivered than any fuel in use in the nineteenth century. What could that be? In the late nineteenth century, the Scottish physicist Lord Kelvin thought he had the answer: gravitational energy.

It is regrettable that we come to Lord Kelvin in one of his errors, for he is well known for many things that he got right, including his work on thermodynamics, especially his role in the development of the absolute scale for temperature and the law of conservation of energy. However, even his wrong answer about the power source of the Sun plays an important role in understanding how the Sun's burning started.

We know that it takes work to lift a heavy object, and we can get a heavy object to do work for us in the process of lowering it. (The counterweight system of an elevator uses this principle. As the elevator is raised, the counterweight is lowered. The energy to raise the elevator is not supplied by the motor, but rather by the lowering of the counterweight.) Similarly, if the Sun formed from a widely dispersed cloud of gas, then in the process of the gas cloud shrinking under the action of its own gravity, work would be done and the gas would be heated. This heating of a body that contracts under its own gravity is called the Kelvin-Helmholtz mechanism. If a widely dispersed gas cloud with the mass of the Sun collapsed to an object the size of the Sun and with uniform density, the amount of energy released as heat per kilogram is about a thousand times as much as for natural gas, which would be enough to keep the Sun shining for about 10 million years. This time scale is much longer than the time scale if the Sun was burning conventional fuels. However, it is still uncomfortably short by geological and evolutionary standards. Indeed, Lord Kelvin thought he had found a fundamental difficulty with both geology and evolution and had a long-standing argument with Darwin about this issue.

This kind of conflict between branches of the sciences or within a branch has good and bad effects. The bad, of course, is that when people argue, they tend to become stubborn and stick to their positions more than they might if they could work things out on their own. Scientists can be as prone to this vice as anyone else, but they have an outlet that counters the bad. If scientists are fighting over the facts, they can always try to demonstrate that their position is correct by creating experiments that test their own and the other side's hypotheses. If the tests show one way or the other, then (in science at least) one side should concede and accept what has been shown.

Sometimes, however, the answer comes from neither side of the argument but from somewhere else, and neither Kelvin, the geologists, nor the biologists were responsible for resolving this dispute. That answer came from a branch of physics that did not yet exist at the time of this controversy: nuclear physics.

It turns out that the Kelvin-Helmholtz mechanism does not explain what powers the Sun. Rather, the Sun is powered by nuclear fusion. The Kelvin-Helmholtz mechanism is simply the spark that gets the fusion started. Fusion as the source of the

Sun's power was first suggested in the 1920s by the British astrophysicist Sir Arthur Eddington, who is best known for his measurements of the bending of starlight by the Sun that confirmed Einstein's theory of general relativity. The details of the nuclear reactions that power the Sun and stars were worked out in 1938 by the physicist Hans Bethe. Bethe was one of the pioneers of what was then the new field of nuclear physics. His papers on the subject were known as the "Bethe bible" and were read by all serious researchers in the subject. In the Manhattan Project, Bethe was the head of the theory division, where he supervised the young Richard Feynman.

But what is fusion? Chemistry looks at atoms as the building blocks of molecules because chemical reactions involve the combining of atoms into complex structures. Atoms combine by sharing electrons. We don't want to dig too far into the classical chemistry of this, which explains the formation of molecules, nor the quantum chemistry, which explains exactly how electrons are shared, but we will offer one of the way-too-simplified explanations and suggest further reading in chemistry. Every atom has spaces for electrons at the different possible energy states (these "spaces" are organized into orbits and orbitals). If an atom has certain unfilled spaces, it can share electrons with another atom. The exact conditions for this require more detail than we want to go into, but in essence one or more electrons become electrons of more than one atom. This binds the atoms closely together and creates a whole new level of structure: the molecule. Some atoms have a lot of spaces to share (carbon is one such) and can create extremely complex molecules by sharing with more than one other atom. Carbon in particular can share with up to four other atoms, allowing it to make long chains of carbons, each of which can have other atoms sticking off it. These carbon chains are the essential structures of life.

But enough of that for now, because, truth to tell, we were only talking about this to say this was not what we were talking about. Because in chemistry, atoms themselves never change. Hydrogen is hydrogen, helium is helium, carbon is carbon, and so on. No chemical interaction can alter the nature and character of an atom because chemical interactions never change the nucleus.

But there are other kinds of interactions, atomic and subatomic ones—reactions that occur not in the electron shells around the atoms but in the nuclei that form their centers. Nuclear fusion involves the combining of the nuclei of atoms so that a new atom is created from the old ones and therefore a new element comes to be. There is also nuclear fission, which splits up large atoms into smaller ones. These nuclear reactions are infamously more powerful than chemical reactions. A kilogram of hydrogen fused into helium provides about 10 million times as much energy as a kilogram of natural gas does when it is burned. The energy from fusing all the hydrogen in the Sun would be enough to keep it shining at the present rate for about 100 billion years (though we will see later that the Sun will consume only about 10 percent of its hydrogen during its lifetime).

Let us look again at the structure of atoms. We noted before that atoms are made of nuclei and electrons, and that chemistry is the interaction of atoms through sharing their electrons. We paid little heed to the nuclei, but now they crave our attention. What, then, are nuclei made of? The basic constituents are two kinds of particles called protons and neutrons. A proton, which alone can be the nucleus of a hydrogen atom, is almost two thousand times more massive than an electron and has an opposite electrical charge to the electron, so that an atom of hydrogen is electrically neutral (charge of +1 from the proton + charge of −1 from the electron = charge 0). A neutron has zero electrical charge and is slightly more massive than a proton. Opposite electrical charges attract, and it is the attraction between electrons and protons that holds an atom together. This attraction between electrons and protons is called the electromagnetic force. Under this force, opposite charges attract each other (positive attracts negative, negative attracts positive), but same charges repel each other (positive repels positive, you get the idea). Therefore, due to the electromagnetic force, the protons in a nucleus (which are positively charged) repel each other, so what holds the nucleus together?

There is a force in nuclei that acts contrary to the electromagnetic force, holding the nuclei together. This force is called (dramatic drum roll): the strong force. Over very short distances, the strong force is stronger than the electromagnetic force, and it is the strong force attraction between the nucleons (a short-form word that just means protons and neutrons) in a nucleus that holds the nucleus together (as long as there are not so many protons as to overwhelm the strong force and break up the atom). The strength of the strong force is connected to the energy released if strong-force bonds are forged or broken. The stronger a force, the more tightly it binds things together, the more energy is contained in its bonds. Chemical energy is the energy needed to bond atoms together in the sharing of electrons; the force there is electromagnetic, which as we noted is weaker than the strong force. It is because the strong force is much stronger than the electromagnetic force that nuclear energies are much larger than chemical energies.

Physicists are known for coming up with whimsical names, and at first sight it seems that "strong force" is just one of those names; why not call it the "nuclear force"? To answer this question, we apply an old line from the comedian Henny Youngman, who when asked "How's your wife?" would reply "Compared to what?" One can similarly ask about the strong force, "Strong in comparison to what?" It turns out that there are two different nuclear forces: the strong force and a much weaker one known as (tepid drum roll): the weak force. The names "strong force" and "weak force," then, make sense in comparison to each other and can be thought of as abbreviated versions of "strong nuclear force" and "weak nuclear force."

In addition, it turns out that all the forces we encounter in daily life (except for gravity) are consequences of the chemical forces between atoms, which in turn are consequences of the electromagnetic force. Pushing and shoving, jumping, dancing, sneezing, and so on—all the day-to-day forces are at the root electromagnetic, because they consist of molecules affecting other molecules (like the molecules in a hammer affecting the molecules in your thumb). The electromagnetic force is weaker than the strong force and stronger than the weak force. Since the electromagnetic force is the one we encounter in our daily life, we can think of its strength as being the normal strength for a force. By comparison with this "normal" strength force, the strong force is strong and the weak force is weak. (Gravity is even weaker than the weak force, so if we were to push this terminology even further, we would have to call gravity the ultraweak force, but gravity was named long before these others, so it gets to keep its old name, which comes from the Latin word for heavy).

Why don't we notice the strong force in our everyday macroscopic lives? The strong force is only strong at a very short range. A nucleus is much smaller than an atom, and when nucleons are much farther apart than the size of a nucleus, the strong force becomes so small as to be negligible. So we could call it "strong over a very short distance, but larger than that who's gonna notice" force, but that might be a little unwieldy.

The process that powers the Sun (remember that's what we're talking about) is a set of nuclear reactions whose net effect is to take four hydrogen atoms and through strong-force and weak-force interactions produce a helium atom. A nucleus of helium consists of two protons and two neutrons bound together by the strong force. We can get energy by combining two protons and two neutrons together to make helium. In the Sun there are plenty of protons since there is plenty of hydrogen and the proton is the nucleus of hydrogen. However, there are no neutrons, except those already bound in nuclei. How then does the Sun make helium?

This question whose origins lie in astronomy is answered within particle physics, by answering the question "Where do neutrons come from?" Particle physics is the branch of physics that concerns itself with the ultimate answer to the question "What are things made of?" A preliminary answer to this question is that things are made of atoms. The word "atom" comes from the Greek word *atomos*, meaning indivisible. But naming something "indivisible" is nearly as dangerous as calling a ship "unsinkable" (remember the *Titanic*?). You are almost daring nature to prove you wrong. Sure enough, the atom is not indivisible, being made of protons, neutrons, and electrons. Since the name "atom" is already taken, particle physicists call the things that are not made of anything smaller "elementary particles." Are protons, neutrons, and electrons elementary particles? Electrons seem to be; but protons and neutrons are made of smaller particles called quarks. The photon is also an elementary particle, and it turns out that the electromagnetic force is a consequence of the exchange of photons among elementary particles; so, for example, two electrons repel each other because one emits a photon that the other absorbs. In this way, the theory of elementary particles, and thus of all matter, also becomes the theory of all forces. Each force has a special kind particle that is passed back and forth in the interactions of the force; these are called exchange particles. The exchange particle for the electromagnetic force is the photon, so electromagnetic interactions use light. This may sound abstruse, but two of the direct applications of this are radio and television—sometimes the abstract isn't as far away as people think.

One of the fundamental laws of particle physics is that for each type of elementary particle there is another type of elementary particle, called its antiparticle, that has the same mass and opposite charge. The antiparticle of the electron is called the positron. The photon is its own antiparticle. The current state of the art in particle physics is called the standard model, which accounts for all known matter, except the dark matter and dark energy to be discussed later, and all known forces, except gravity. The standard model is one of the great triumphs of twentieth-century physics.

What does all this have to do with making neutrons? It turns out that there is an elementary particle called a neutrino, with no electric charge and very small mass, and that given enough energy, a proton can turn into a neutron, positron, and neutrino. This kind of particle equation works both ways: a proton, if given enough energy, can become the set of three particles given above, and a neutron with a positron and a neutrino can become a proton. A neutron has more mass than a proton—in fact, it has more mass than a proton and an electron together. It takes energy to make mass, so protons left to themselves will not turn into neutrons, but with assistance they can do so.

This energy can be provided by another proton combining with the neutron to produce deuterium, a heavier form of hydrogen. A deuterium atom has as its nucleus one proton and one neutron. It is chemically the same as hydrogen, but is about twice as massive. Two chemicals with the same number of protons but different numbers of neutrons are called isotopes, so deuterium is said to be an isotope of hydrogen. Isotopes behave the same as each other chemically since chemical reactions depend on the number of electrons orbiting an atom, and the number of electrons depends on the number of protons. Chemically, neutrons do not matter. In the nuclear arena, however, neutrons pack a heavy punch.

This first step in the Sun's fusion reaction is that two protons combine to form deuterium, a positron, and a neutrino. This sounds like a complicated way to make deuterium: it would be much simpler to just combine a proton with a neutron. However, while the Sun contains protons by themselves, it does not contain neutrons by themselves. This is because a neutron by itself turns into a proton, an electron, and an antineutrino (the antineutrino is the antiparticle of the neutrino just as the positron is the antiparticle of the electron). This sort of spontaneous transformation of a particle left to itself is called decay. On the average, a neutron by itself will decay in about fourteen minutes.

Neutrons in some but not all nuclei decay. Those nuclei in which this occurs are called radioactive, and the decay is called beta decay ("beta" in this case refers to the emission of an electron called a beta ray in this context). In such cases, the atom changes its chemical properties, since a neutron's decay leads to a jump up of one in atomic number (the number of protons), hence an atom of radioactive gold (atomic number 79) undergoes beta decay into an atom of mercury (atomic number 80).

There are other ways in which a nucleus can decay: alpha decay (the nucleus ejects a helium nucleus, which in this context is called an alpha particle) and gamma decay (the nucleus ejects a highly energetic photon, which in this context is called a gamma ray). Decays only happen spontaneously when the resulting atom has less energy than the starting one. The entire field of radioactivity revolves around this kind of decay, and there are many applications of the study of it, from the dating of archaeological finds to the treatment of cancers.

Fusion today is not just a theoretical process. It has been employed in the most dangerous weapons we have. Hydrogen bombs use a process similar to that in the Sun to release energy with horrifying efficiency. In bombs the process is the fusion of two different kinds of hydrogen: deuterium (one proton and one neutron) and tritium (one proton and two neutrons). This reaction is both powerful and fast. If the Sun used this kind of fusion, it would burn out even faster than the equivalent mass of fossil fuels.

The reaction in the Sun must be powerful but slow for the Sun to have lasted as long as it has. The reaction that turns a proton into a neutron proceeds through the weak force, which is weaker than the electromagnetic force. In quantum mechanics, the weaker a force is, the less likely it is for a reaction that takes place through that force to occur. Therefore, weaker reactions are rarer and hence processes that rely on weak reactions take longer on the average than those that have strong reactions. For the slowness of solar fusion, the critical link in the chain is the reaction that produces a neutrino. This is the first and slowest of the steps in the Sun's fusion reaction. Neutrinos interact only through the weak force. This means that a reaction involving neutrinos, including the one in solar fusion, is very unlikely to occur.

At the center of the Sun, temperatures are very high and protons encounter each other frequently. However, it is rare that such an encounter both overcomes the electric repulsion of the protons for each other and undergoes the weak interaction that produces the neutrino. In other words, because neutrino-producing interactions are uncommon, solar fusion is kept at a relative trickle. This same weakness of interaction means that the neutrinos that are produced in the center of the Sun readily emerge from the Sun without being impeded by the enormous amount of matter between the center and the surface.

The last several paragraphs describe things that are all in the theoretical universe, although we baldly stated them as fact. Right now, we need to come to a screeching halt and ask how we know whether any of this is right. After all, we are talking about reactions that occur at temperatures of millions of degrees at the center of the Sun. We can't design a space probe that would survive those conditions. So why is this science rather than science fiction?

Or more briefly, "How do we know that?"

Part of the answer comes from solar models, which while also theoretical have more detailed aspects that can be put to the test. Solar models are detailed theories of the Sun where the Sun is considered for theoretical purposes as a set of thin spherical layers, with the first layer at the center and the last layer at the surface. The temperature, density, and pressure are given for each layer of the Sun by these models. These models, to fit detected and theoretical knowledge, are constrained by several considerations that arise from experiments and analysis in different environments:

1. Hydrostatic equilibrium: in each layer of the Sun, the pressure pushing that layer outward must balance the force of gravity pulling it inward. (The word "hydrostatic" shows the origin of this idea in the study of water.)
2. Energy transport: in each layer the energy coming from below plus the energy produced in the layer must equal the energy that comes out of the layer. In other words, at each layer of the Sun, the energy coming out must be what comes into the layer, plus what is made there.
3. Nuclear reactions: the nuclear reactions in the Sun must occur at the rates at which they have been measured to occur in the laboratory, since they are the same nuclear reactions.

Just as solar spectra must be the same as chemical spectra, so solar nuclear reactions have to be the same as terrestrial nuclear reactions.

4. And finally: the model must give the correct mass, power, and surface temperature for the Sun, and these, as we have shown, are determinable quantities.

There happens to be such a model. The fact that there is one is a (partial) confirmation of the theory, since a model that works for what has already been observed is a model that can be adopted and tested for later experiments. Since the solar model (the layers given above and energy powered by the listed fusion reactions) does a good job, we can speak with reasonable confidence about things happening at temperatures of millions of degrees deep below the surface of the Sun.

But we are not stuck with testing theories against already observed results. Remember that the above description of solar power says that a huge number of neutrinos are emerging from the Sun. If we can find a way to test whether there is such an effusion, we will have more confirmation of the theory. Since neutrinos interact so weakly with matter, they pass readily through the whole Sun. By the given theory, the neutrinos from the Sun are coming directly from its center. If we detect them, we have a direct probe of conditions in the Sun's center. However, the wonderful property of neutrinos that they are a direct piece of information about the center of the Sun also makes them an experimentalist's nightmare. In order to detect them, we must make an experimental apparatus that the neutrinos do not simply pass through without hitting anything. Neutrinos as particles have no charge, very little mass, and the only thing they do in the universe is occasionally interact with other particles weakly. In short, if neutrinos can readily pass through the whole Sun, what hope do we have of making anything that can stop them?

We have hope. The predictions of quantum mechanics, which govern all subatomic particles, are always in terms of probabilities. Instead of saying, "This particle will collide with that one producing this other particle," quantum mechanics says, "There is a (ridiculously small number that you would never want to bet on even if you regularly play the lottery) chance that these two particles will interact producing this other particle." Thus quantum mechanics does not say that neutrinos do not collide with and therefore interact with atoms, just that an individual neutrino has a very small probability of hitting an individual atom. This probability is so small that a given neutrino is more likely than not to pass through the entire Sun without hitting any of the atoms in its path. It is therefore certainly true that in any experimental apparatus we build to detect them, most of the neutrinos will simply pass through the apparatus without being detected. But some tiny fraction of the neutrinos will hit an atom in the apparatus, and we can then reason from the number detected and the small probability of interaction to find the number of neutrinos emitted by the Sun.

We do this using an idea in probability theory called expectation value. If you roll a hundred dice, you expect to get a number of sixes nearly equal to the number of dice (100) times the probability of rolling a six on a given die (1/6), which gives an expectation of 50/3. You can turn this around and check how many sixes you got and from that estimate how many dice you were rolling. The neutrino experiment behaves the same way: each neutrino is a die, and quantum mechanics tells us the odds of interaction. We then need to measure the number of interactions and from that estimate the number of neutrinos.

The larger the apparatus is, the more likely it is that a neutrino will hit an atom in it. So the key properties of a solar neutrino detector are that it must be large and that it must be shielded from anything else that might cause the atoms to change. The best way on Earth to do this is to put the detector deep underground to shield it from anything other than neutrinos. The first solar neutrino detector, built by American physicist Ray Davis in 1967, consisted of 100,000 gallons of cleaning fluid in the Homestake Gold Mine in South Dakota, almost one mile below ground. The cleaning fluid contained chlorine, which when hit by a neutrino turns into a radioactive isotope of argon. The experiment then consisted of counting how many atoms of argon were produced. The Davis experiment found one argon atom produced every two to three days.

One atom in 100,000 gallons sounds impressive, but is even more impressive when one considers how small atoms are. Depending on what type of atom it is, the diameter of an atom is from about one to a few ten-billionths of a meter. Or to put it another way, there are more atoms in a drop of water than there are drops of water in an ocean. Searching for a single argon atom in a 100,000-gallon tank of cleaning fluid is much worse than looking for a needle in a haystack. Every two to three months, Davis and his team would bubble helium through the cleaning fluid and then circulate the resulting gas through an elaborate apparatus that trapped any argon atoms in charcoal. The charcoal was then removed and placed in a device called a proportional counter. The argon atoms were radioactive and would eventually decay back to chlorine, with the decay process ejecting an electron from the resulting chlorine atom. The proportional counter would produce a signal when it detected each of these electrons.

In the field of experimental design, this brilliantly simple device is wonderful because of how the theory led to the creation of such a simple, if bulky, apparatus. By combining the requirements of depth and size, Davis took an extremely easy-to-find material (chlorine) and used it to detect some of the most elusive objects in the universe. This kind of connection between the tiers of the universe is the pathway through which science grows. It is worth reiteration: Theory leads to design for detection; detection leads to perception; perception is codified in detection, which confirms or refutes theory. That is the basic cycle of science, seen here in a giant vat of cleaning fluid blipping out occasional argon atoms.

The detection of neutrinos from the Sun was a triumph. However, an important discrepancy remained. The number of neutrinos detected by the Davis experiment was only about one-third of the number predicted by the solar models. This discrepancy came to be known as the solar neutrino problem. In these days of inaccurate models for everything from the weather to the stock market, it is natural to suppose that this discrepancy is simply due to inaccuracy in the model of the Sun. Indeed, some astronomers and physicists thought this and much work was put into a careful examination of the solar model to see whether it needed to be modified.

One subtle point is that the neutrinos detected by the Davis experiment come not from the first step of the main fusion reaction discussed earlier, but instead from a different reaction that occurs more rarely. Thus it was conceivable that the solar model could be consistent with the power produced by the Sun but nonetheless got the number of neutrinos wrong. However, decades of careful examination of the solar model yielded the same answer. The model predicted about three times as many neutrinos as the Davis experiment detected.

The solar neutrino problem was finally resolved only in recent years by a variation on the Davis experiment called the Sudbury Neutrino Observatory (or SNO, pronounced "snow," one of many whimsical acronyms that physicists are known for). Like the Davis experiment, SNO is built deep underground in an abandoned mine. However, SNO uses heavy water (about 1,000 tons of it) instead of cleaning fluid. "Heavy water" sounds like an odd name, but adding a neutron to the nucleus of any atom makes the atom heavier but doesn't change its chemical properties. Deuterium (hydrogen with an extra neutron) is therefore well described as heavy hydrogen. Water (H_2O) is a combination of hydrogen and oxygen; and heavy water is simply water made with heavy hydrogen. It is chemically the same as ordinary water, but is a little heavier because each hydrogen atom has an extra neutron.

In the Davis experiment, the neutrino turned chlorine into argon by turning a neutron into a proton and an electron. But it had long been known that there are three different types of neutrinos. The one we have been discussing is called the electron neutrino, and the other two are called the muon neutrino and the tau neutrino. The explanation for these three kinds of neutrinos is deep in particle physics, where there are three separate "families" of particles. Each family has its own particle types, and each family has its own kind of neutrino (the shy kid who doesn't talk to anybody). One of these families consists of electrons, electron neutrinos, and the quarks that make up protons and neutrons.

The other two families don't concern us much here, but the experiments we are talking about both draw upon and confirm several different branches of physics. This is important to understand, because while for purposes of our own understanding we humans divide the sciences into branches based on subject matter, the universe does not so divide itself. The different structural levels of the universe all rely on each other. Just as the sound of an orchestra can be defined or ruined by the playing of the second fiddle, so the power of stars relies on the properties of neutrinos.

For our purposes, what is important about these families is that electron neutrinos can turn a neutron into a proton and an electron, but the other two kinds of neutrinos cannot. It was therefore recognized that if some of the electron neutrinos produced in the center of the Sun turn into muon neutrinos or tau neutrinos before they reach the detector, then this could resolve the solar neutrino problem (since if only one-third of the neutrinos that emerged from the Sun were the right kind, then we would expect to detect only one-third of the number of neutrinos that we originally expected to detect, which is just what the Davis experiment did).

To see whether this explanation is correct, it is necessary to have a detector that can detect all three types of neutrino. This is where the heavy water comes in. Remember that the deuterium in heavy water is a proton and a neutron stuck together. Any type of neutrino can collide with deuterium and (if it has enough energy) break it apart into a proton and neutron. The neutron will then decay into a proton, electron, and antineutrino, and the electron can then be detected by the light it gives off as it goes through the water.

The result of the SNO experiment is that the number of neutrinos detected agrees with what is predicted by the solar models. By this conjunction of theories and experiments, the solar neutrino problem has been solved and we can at last be decently confident that we know the source of the power of the Sun. This allows confidence in information relevant to the age of the Sun. (There is a kind of domino theory in logic, if *A* depends on *B*, which depends on *C*, which depends on *D*, then if you show *D* to be true, then *C*, *B*, and *A* will be shown to be true.) At the current rate of power consumption, the Sun would take about 100 billion years to use up all its hydrogen. Note that this doesn't directly tell us how long the Sun has been burning or how long it will last. The answer to the last question depends on how much of its fuel the Sun will eventually consume, a topic to be treated in the next chapter.

Just as neutrinos give us direct information about the center of the Sun, we would like to have direct information about the chemical composition of the Sun rather than having to rely solely on the methods of spectral analysis. If two independent means of observation can come to the same conclusion, the conclusion will be bolstered, as will the theories that lead to the observations.

It turns out that this more direct observation is possible. For the most part, each layer of the Sun is in hydrostatic equilibrium, with gravitational attraction balanced by pressure. However, this is not true for the outer layers (see fig. 4). When we look at the Sun, it appears to have a definite size, as Earth does. However, this is an illusion. The Sun is mostly a ball of plasma (gas where

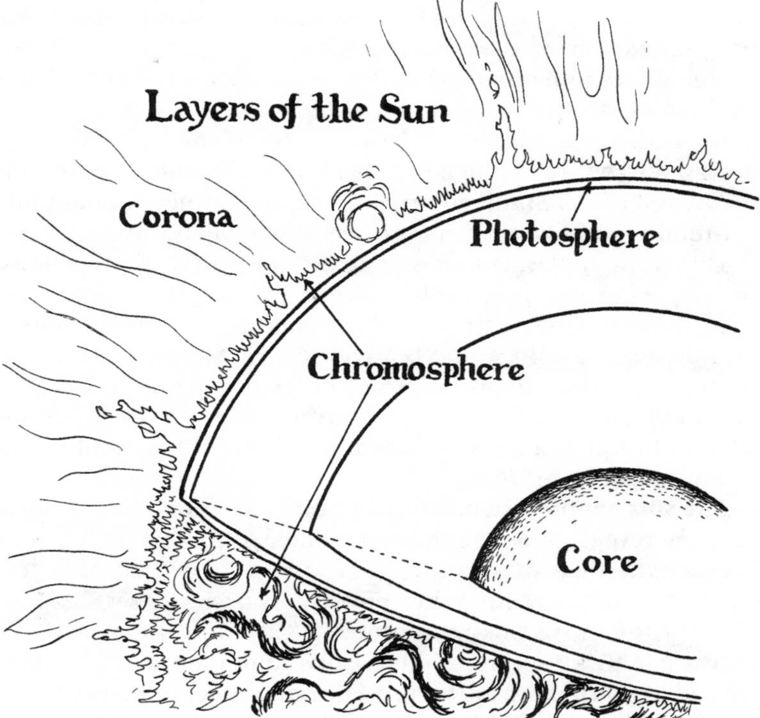

Figure 4

the atoms have lost their electrons). The plasma is most dense at the center and then gets less and less dense the farther from the center one goes.

The Sun seems to have a sharp surface because the light on its way out from the center keeps on bumping into nuclei or electrons. The last place that the light bumps into something before coming to us appears to us to be the place the light is coming from. This last place is the photosphere, the light from the photosphere tends to get away and become the light we see and from which we decide that we have seen the surface of the Sun.

However, the plasma does not stop at the photosphere. There are layers of the Sun above what appears to us as the Sun's surface, solar matter we rarely see. The outermost layer is called the corona, and it is not in hydrostatic equilibrium. Instead, it has so much energy transferred to it by the magnetic fields of the Sun that the Sun's gravity is not sufficient to hold it in place and plasma is steadily flowing away from the Sun. This flow of plasma is called the solar wind. The wind consists mostly of protons and electrons.

How much material is the Sun losing in the solar wind? It turns out to be about 2 billion kilograms per second. That sounds like a lot, but the Sun is so massive that this comes to about one ten-thousandth of a solar mass every billion years.

This wind of plasma blows out from the Sun in all directions, creating a current in our solar system. The empty space around us is not completely empty on the atomic and subatomic levels. And the wind itself is quite fast and strong. Indeed, recently an attempt was made to bring an idea of science fiction into reality by building a spaceship that sailed using the solar wind. Regrettably technical problems thwarted this effort, but other attempts will no doubt be made.

The solar wind is detectable by indirect observation since it is partly responsible for the tails of comets. As the comets pass through the solar wind, they interact with it energetically, producing the appearance of the tail. The solar wind is also responsible for the more earthly (and unearthly) aurora borealis, or northern lights. Protons and electrons (and indeed any particle with electric charge) in the presence of a magnetic field travel in a spiral around the direction of the magnetic field. The only places these charged particles can come to Earth are the poles of that magnetic field; the north and south poles of Earth's magnetic field are close to its north and south geographic poles. When the charged particles hit Earth's atmosphere, they ionize atoms (force electrons to jump away from the atoms) and thus cause them to glow. This glow is the northern lights. The same thing happens in the region around the South Pole, where it is called the aurora australis, or southern lights.

It is in theory possible to catch the solar wind and from it find out what materials are leaving the Sun and hence determine a bit of what the Sun is made of. Recent experiments have attempted to do this. The solar wind is a case where the usual hydrostatic balance between pressure and gravity is broken. In this case, pressure wins and material flows outward, slowly diminishing the Sun.

In the next chapter we will consider what happens instead when gravity wins and a star collapses to form a black hole—for although we have looked at and learned much about the Sun itself, it is not alone. Close to us and vital to our lives as it is, the Sun is but one example of a star, and to widen our understanding, we must take from it its favored place in the universe. We will use our comprehension of the Sun as a first step out into the wider cosmos, and in so doing we will place it within a much broader context, in the space of the galaxies and the time of stellar evolution where it is but one among many and by no means exceptional.

We will look at the life and death of stars, in particular the unnervingly simple stellar fate of becoming a black hole.

Step 2

BLACK HOLES

Previous page: Black hole jets. NASA/CXC/CfA/R. Kraft et al. 2008.

CHAPTER 4

Old Gravity and New Gravity

While none have ever been perceived, black holes loom large (okay, massive) in the detected and theoretical universes. Black holes and their effects have been studied using a wide variety of detection instruments: the usual optical telescopes as well as radio telescopes and more exotic satellites that act as telescopes for ultraviolet light, X-rays, and gamma rays. And now the first steps are being taken to study black holes by detecting the gravity waves they emit when they collide. Theories about black holes have been made by an even wider variety of theorists: Astrophysicists study them to find the ultimate fate of massive stars and the nature of the enormous amounts of energy pouring from the centers of some galaxies. General relativists, physicists who specialize in Einstein's general theory of relativity, see black holes as one of the most interesting predictions of that theory and hope that further study of black holes will tell them more about the nature and implications of general relativity. Mathematicians find that black holes provide an application for the arcane techniques of differential geometry, the study of curved spaces. Particle physicists, string theorists, and others interested in combining general relativity with quantum mechanics find the quantum behavior of black holes one of the most fruitful areas for this research. These observations and theories show black holes to be both some of the most exotic and at the same time some of the simplest objects in the universe.

In order to understand black holes, we need to look at what they were before they were black holes, so we must first understand stars. By sheer coincidence (not that we planned this at all), this understanding can be built from what we have already come to understand about the Sun. Not only will we use the information we gleaned in the previous section; we will also use the same methods, albeit in a different direction.

Previously, we took apart the perceived Sun and put it back together again using theory and detection, so that the Sun we now know is more than the Sun we simply see. We also by implication took apart and put back together stars, although why we can justly say that the analysis of the Sun applies to stars will have to wait until later in this chapter.

The process of dissection and reconstruction began with a perceived object from which we, using theory and detection, came to understand it. This time we are going to begin with an object that was at first only theoretical and see whether or not we can detect something that fits the theory. This is a vital test of the three-tiered universe, whether or not theory and detection can uncover unseen objects, whether using two tiers of our universe can find unseen objects in the universe of fact. This is the opposite direction of science from that used with the Sun. Before, we saw and did not understand a thing, so we used science to find understanding; now we seek to take understanding and from it seek to know if there is a thing to be understood. Instead of climbing up from perception to detection to theory, here we climb down the ladder from the abstraction of the theoretical universe to the everyday of the perceived to find what cannot be seen and yet is.

The theory we will begin with is that of gravity. However, we will not employ the elaborate modern theory of gravity begun by Einstein until later. Initially, we will stick with gravity as Newton conceived of it, as a force that attracts objects to each other. This is the theory of gravity we outlined in the previous section that we used to discern many properties of the Sun.

We start our current project with the detected fact that the Sun and other stars are massive objects with consequently huge gravitational attraction. We add that to the bit of Newton's theory that says that gravity affects all objects, and put in an assumption that this effect of gravity applies to light as well. We then combine these bits of detection and theory to create a theoretical question. One of the vital steps in science is the process of combining detections and theories to create questions. These questions are then used as guides to more work in theory and the creation of appropriate experiments in order to answer the questions.

In many endeavors, it is the framing of the questions that matters. Questions serve as guides to the mind; they create expectations for the form of answers, shaping as yet unknown thoughts. Consider the difference between the questions "Who did this?" "What did this?" and "How did this happen?" The first question creates the expectation of a person who was the cause of something. The second question still expects a cause as answer, but now the cause need not be a person. The third question expects a process as an answer. To elaborate the example backward:

"How did the cookie jar break?"

"It fell off the shelf onto the floor."

"What broke the cookie jar?"

"The floor did."

"Who broke the cookie jar?"

"Um, uh, I did. Sorry."

In this case, we will use Newton's theory of gravity to create the question of whether there can be a star whose gravity is so strong that the light cannot escape from it. Is there a star so massive that it cannot be seen?

To answer this, we grab for the most relevant concept that might be of use: escape velocity (last seen floating away in a helium balloon). Newton discovered that since the force of Earth's gravity gets weaker the farther up one goes, an object at Earth's surface that is moving sufficiently fast away from that surface will be able to escape from Earth's gravity. How fast does it need to be moving in order to transcend the earthly and soar into the heavens? (Sorry about that, but there's something poetic about escape velocity.) From Newton's formula for the force of gravity and a few quick calculations, the escape velocity can be determined. The escape velocity for Earth is about 11 kilometers per second, much faster than we can throw a baseball, but not, as we've seen, beyond our rocket technology. For the Moon, the escape velocity is about 2.4 kilometers per second.

This tells us why Earth has an atmosphere and the Moon does not. In the last chapter we explained the absence of helium in Earth's atmosphere by the ease with which such a light atom could attain escape velocity. In the lesser gravity of the Moon, much heavier atoms and molecules can escape easily. Any gas the Moon once had would long ago have escaped, since the gas would simply become hot enough from solar energy to fly away.

Using the calculation of escape velocity and using the speed of light, we can calculate a formula that given the mass of an object tells us the radius that an object needs to have for its escape velocity to equal the speed of light. The formula for the radius is

$$R = 2GM/c^2$$

Here, R is the needed radius, G is Newton's gravitational constant, M is the mass of the object, and c is the speed of light. Notice that we have written this formula in such a way as to yield the necessary radius if we have the given mass. We have done this because if you think about it, any mass no matter how tiny could in theory become such a "black" object if it were only small enough in size, since gravitational force depends both on mass and distance. We could equally well turn the formula around to yield the mass that an object of a given radius would need to have in order to be "black." In fact, we will find it useful to ask a related question: Given a certain density (mass per unit volume), how much mass at this density do we need to make a black object? The radius formula given above can also be used to answer this question. This is more sleight of formula where we take two different formulas that talk about the same quantity and combine the two to produce new awareness of that quantity.

The radius given by the above equation is called the Schwarzschild radius, named for Karl Schwarzschild (1873–1916), a German astronomer who discovered the solution of Einstein's theory for any spherical object. Schwarzschild made this discovery just at the end of his life (he died of an illness contracted while serving in the German army in World War I) and just shortly after Einstein published his general theory of relativity. Einstein was surprised that an exact solution of his theory could be found and that it turned out to be so simple.

We can use this formula in at least two different ways. For a given mass, we can ask what the Schwarzschild radius is, or for a given density of matter, we can also figure out how much mass we need at that density so that the radius of the resulting object is its Schwarzschild radius. It may seem that treating all objects as if they were spheres with radii, even though the universe is full of a diversity of shapes, is an oversimplification. But given time, the force of gravity tends to make massive objects spherical (at least if they are not rotating too fast). So far as can be determined, we don't lose much generality talking about spheres, and later on we will broaden the discussion to treat rotating black holes, so that covers most of the bases.

Let us ask both of the above questions for the Sun. What is its Schwarzschild radius? Calculation yields about 3 kilometers. So if the Sun were only 3 kilometers in radius (6 kilometers across) as opposed to being about 700,000 kilometers in radius, it would be such a black object. Oh, let's just call them black holes. That's what we were going to do anyway. And what mass would an object need at the Sun's density to be a black hole? The Sun's density is about that of water, so again after some calculation, we find the hypothetical object would need to be about 100 million solar masses (that is 100 million times the mass of the Sun).

The above discussion may sound like a modern bit of theory, since it is commonly believed that black holes are the province of present-day astrophysicists working with computers and space-borne telescopes. But the question asked and the answers given above require only Newton's theories and algebraic calculations that can be done by hand. Indeed, all of the above were done in 1783 by John Michell and in 1796 by Pierre-Simon Marquis de Laplace. Michell incidentally has already shown up invisibly in our discussions. He invented the torsion balance used by Cavendish to weigh Earth. Laplace (1749–1827) was a French mathematician, astronomer, and physicist. He is one of those people whose influence in math and the sciences is broad and deep. He is known for showing that the solar system is stable. He did extensive studies of the motion of the planets, taking into account not only the gravitational effects of the Sun but also the gravitational effects that each planet has on the others. Two mathematical objects are named after him: Laplace's equation, used in problems of electrostatics, and the Laplace transform, used in differential equations. Laplace served for six weeks as

Minister of the Interior under Napoléon, who was not impressed with Laplace's performance, saying that he "sought subtleties everywhere, had only doubtful ideas, and carried the spirit of the infinitely small into administration." (It is not known to us what Laplace thought of Napoléon's record as a mathematician.)

Laplace and Michell worked on black holes as a theoretical concept more than two hundred years ago, but they faded from sight as only intellectual curiosities (no, this is not the last "can't see them because they're black holes" joke—get used to them). Black holes reappeared as consequences of the theory of general relativity, but even then they remained purely theoretical for some time. How black holes were finally detected we will talk about later. For now, we'll continue our stay in the theoretical universe.

We now pause for a bit of concern. Newton's theory of gravity has been supplanted by Einstein's. Ironically, we have to ask the question whether the classical concept of black holes can survive in modern theory. Newton's conception of gravity works very well for objects that move slowly compared to the speed of light and for gravitational fields that are not too strong. But fast-moving objects (of which light is the fastest) are described using Einstein's special theory of relativity, and strong gravitational fields (to be strong enough to keep light from escaping a gravitational field must surely be very strong) are described using Einstein's general theory of relativity. The formulas of Michell and Laplace for the radius of an object that can trap light are derived using a supplanted theory and must be rechecked in Einstein's theory before they can be trusted.

A brief digression on the replacement of Newton's view of the universe with Einstein's is necessary at this stage.

Newton's theory of gravity was tested and found to be true over and over again, and held its own for centuries. But this was only because the limits of Newton's theory—the conditions of high speed and gravity where it failed—were beyond the ability of scientists to detect. The process by which Newton's theory of gravity was replaced by Einstein's began in the nineteenth century with an innocent-sounding question: How fast are we moving through the medium that propagates light? Just as ocean waves are a disturbance of the normal height of the water and sound waves are a disturbance of the normal pressure of the air, so light waves were presumed in the nineteenth century to be some sort of disturbance of some sort of medium that was called the ether.

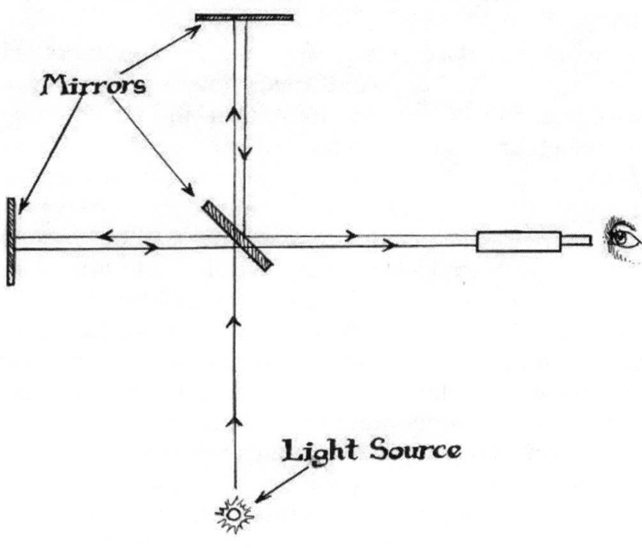

Figure 5

This is one of those cases, as we shall see, where the question itself misleads. "How fast are we moving through the ether?" was a natural question to ask. An experiment to answer that question was done in 1887 by two Americans: the physicist Albert Michelson and the chemist Edward Morley, two names now bound together forever in the annals of science. The experiment used an ingenious L-shaped device that Michelson had invented called the interferometer (see fig. 5). At the corner of the L, in front of a light source, is a mirror whose reflective coating is so thin that it reflects half the light and lets the other half pass straight through. (This half-reflective mirror is called a beam splitter.) The effect of this splitter is to break the beam into two beams, each going down one arm of the L. Each beam travels down its arm and is reflected by the mirror at the end of that arm, and then the two beams recombine at the corner of the L.

If the two beams take the same amount of time in their trips, then the beams when they recombine reinforce each other to make a bright beam. However, if one beam's travel time is longer than the other one's by half the period of the light wave, then the two beams work against each other in a wave phenomenon called destructive interference and the combined light beam is *completely dark*. This is easy to understand thinking of light as a wave. If the beams are off by half a period, then the top of one wave is meeting the bottom of the next wave and adding together to nothing. The experiment was based on the notion that if light is moving through the ether at speed c and we are moving in the same direction as the light at some speed v, then we should measure the light as moving relative to us at the slower speed $c - v$. In other words, when we are chasing after the light, it should seem to us that it is moving slower. Now suppose that one arm is aligned with the direction of travel through the ether. Then its light beam goes "upstream" to its mirror and "downstream" to come back, while the other light beam travels "across the current." Michelson and Morley calculated that the two beams would then have different round-trip times and would therefore interfere. Change the orientation of the apparatus and the amount of interference would change. By measuring the amount that the interference changed as the apparatus was rotated, Michelson and Morley hoped to use this measurement to calculate the speed of Earth through the ether.

The result of the Michelson-Morley experiment was puzzling. Rotation of the apparatus produced *no change at all* in the interference of the light beams. This meant that we are not moving through the ether, that the speed of light that we measure is always c. At first one might suppose that by some strange coincidence Earth is at rest in the ether. But this can't be the case because Earth is in orbit around the Sun and therefore has different velocities at different times. The Michelson-Morley experiment was done at several different times of the year but always yielded the same result. Faced with such a surprising result, one might suppose that perhaps the experiment had been done incorrectly. But Michelson had devoted his life to being the best in the world at precision optical measurement. He held the record for the most precise measurement of the speed of light and once measured the distance between two mountain peaks to an accuracy of less than an inch. Doubt other experiments if you like, no one could doubt Michelson: this puzzling experimental result could not be dismissed as erroneous; it had to be explained.

The explanation was provided in 1905 by Einstein in what he then called the theory of relativity. Einstein developed this theory as a grand rethinking of the foundations of Newton's mechanics and Maxwell's theory of electricity and magnetism. In Einstein's theory, there is no ether and all observers measure the speed of light to be *c*.

At first this assertion seems totally crazy. If I chase after a light beam at one-half the speed of light, then surely the distance between me and the beam can only be increasing at the rate of one-half the speed of light. In other words, I should measure the speed of light to be *c*/2. But as the Austrian physicist Wolfgang Pauli, one of the pioneers of quantum mechanics, remarked, "Einstein's theory is not as crazy as it sounds." Pauli was something of an expert on crazy-sounding theories. He is responsible for the Pauli exclusion principle, an essential part of quantum mechanics needed to account for the chemical properties of atoms and to explain why the elements can be arranged in a periodic table. He also first came up with the theory of neutrinos and first theorized that electrons have spin. He was notorious among physicists for his acerbic wit, and there is a legend that he was such a great theorist that all experiments would go wrong in his presence. This is sometimes called the Pauli effect.

Relativity involves a radical modification of the notions of both distance and time. Newton saw the distance between objects and the time things take to happen as the same no matter who was measuring the distance or the time. Einstein discerned that measurements of the same distances and times produce different values if the people making the measurements are moving at different speeds. These differences of measurement paradoxically produce the effect that all observers measure the speed of light to be the same. We will go somewhat more into relativity later, but since it is a big subject deserving of several other books (which fortunately have been written), we defer to those sources.

One consequence of relativity is that nothing can go faster than light. Despite the fact that relativity seems completely against all observation, it has been demonstrated to be correct. Nowadays its effects are detected every day in cosmic ray experiments and in particle accelerators.

Relativity solved a mystery having to do with light (that there is no ether, and that light speed is constant to all observers), but it raised a new one about gravity. The new mystery had to do with an aspect of Newtonian gravity that had disturbed Newton's contemporaries: action at a distance. In Newton's theory, the gravitational force that one object exerts on another depends on the distance between the two objects, and when the distance changes, so does the force. When the Sun moves, its distance from Earth changes instantaneously and therefore, according to Newton, the force that the Sun exerts on Earth changes instantaneously. How, wondered Newton's contemporaries, does the information that the force has changed cross the vast gulf between the Sun and Earth in no time at all?

With the advent of relativity, what was merely disturbing became downright illegal. Action at a distance involves a "signal" (information about the position of the Sun "communicated" to Earth through a change in gravitational force) passing from the Sun to Earth in no time at all and therefore at infinite speed. This violates the prediction of relativity that nothing can travel faster than light.

"So much the worse for relativity," one might retort. But to Einstein it was Newton's theory of gravity that needed to be changed, replaced by a better theory in which all changes in gravity travel no faster than light. It took Einstein ten years to come up with such a theory, and that theory involved a further radical change in the notions of distance and time so that not only did they change with changes in speed, but they changed with changes in the gravitational field. The 1915 theory came to be known as the general theory of relativity, in contrast to the 1905 theory, which came to be known as the special theory of relativity. These names are a bit misleading: what is "special" about special relativity is simply that it applies when gravity is absent or at least sufficiently weak that its effects on space and time can be neglected.

If Einstein's theory of gravity is so radically different from Newton's, and if Einstein's theory is the right one, then why weren't there in the nineteenth century a whole host of detections showing that Newton's theory is wrong? Well, at low speeds and low gravity, the predictions of Einstein's theory are very close to those of Newton's theory. This is a property of any new theory that "corrects" a successful old theory. To be successful, the old theory had to make accurate predictions that were tested by experiment. The new theory, to be successful, must not "throw out the baby with the bathwater." It must, in the cases where the old theory was tested by experiment, make predictions that only differ from those of the old theory by amounts too small to have been found by those experiments. Here "low speed" means low compared to the speed of light, and "low gravity" means a gravitational field for which the planetary orbits are low speed. The speed of light is an enormous 186,000 miles per second, so most familiar phenomena (other than light itself) poke along at low speed. An idea of what low gravity means can be gleaned from the fact that in relativity our entire solar system is low gravity. Experiments were done to verify relativity, but they involved the careful measurement of the tiny differences between the predictions of Newton's theory and Einstein's in the low-gravity environment of the solar system. One such test actually involved detections that had been made before relativity: a tiny discrepancy between the observations of Mercury's orbit and the Newtonian predictions was resolved by the more accurate predictions of Einstein's theory. Another test involved a measurement, made by Eddington during a solar eclipse, of the minuscule amount that starlight is bent by the gravitational field of the Sun.

These effects are so tiny that one might think that general relativity could never have any impact on our daily lives, but that's not true for any of us who have used GPS. The Global Positioning System is a sophisticated array of satellites that each of us can use to find our exact position. Each satellite sends out a radio signal that tells its position and the time when the signal was sent. A GPS receiver, given the signals from at least four of the satellites, can figure out its own position simply using the positions of the satellites and the fact that their radio signals travel at the speed of light. For this system to work, the orbits of the satellites must be known extremely accurately; so accurately that plain old Newtonian gravity is not good enough: general relativity is needed.

What does general relativity tell us about the radius of an object whose escape velocity is the speed of light? To avoid having to present the complicated math of general relativity, we're not going to show you how the answer is found. (Relativity is beautiful in conception but complex in calculation; if you want to follow the math, look in a text on relativity.) The answer turns out to be the same as from Newton: $R = 2GM/c^2$.

At first this seems like an amazing coincidence. But it isn't, because of a simple but powerful technique in physics called dimensional analysis, a very fancy term for a simple but vital idea. The idea is that you always have to know what you're talking about when you measure something. The point of dimensional analysis is that the quantities in physics formulas are not simply numbers, but represent physical quantities that have units. A unit is simply what you mean by the number 1 in a given circumstance.

"How much is there?"

"One."

"One what?" The answer to this question is the unit you are using.

It doesn't make any sense to say that an object has a length of 2, but it does make sense to say that it has a length of 2 meters, because a meter is a unit of length. Similarly, it doesn't make sense to say that an object has a speed of 2 meters, because a meter is a unit of length rather than speed, but you can say that it has a speed of 2 meters per second. The units tell you what kind of thing you are talking about. You measure length in units of length (meters, feet, kilometers, miles, AUs). You measure mass in units of mass (grams, kilograms, solar masses). And so on.

When you add or subtract quantities, you can only do so when they are of the same units. First they have to represent the same quantity. You cannot add 5 meters to 6 kilograms; length cannot be added to mass. You can add 6 to 5 and get 11, but 11 whats? What do you have 11 of? There is no meaningful answer. The operation of adding 5 meters to 6 kilograms is just plain meaningless. You can add 5 meters to 6 kilometers (since they are both units of length), but first you have to convert one to the other. Either you can say 5 meters is .005 kilometers and so end up with 6.005 kilometers, or you can say that 6 kilometers is 6,000 meters and end up with 6,005 meters. They both mean the same thing, only the particular unit of length is different.

To reiterate, addition and subtraction can only be done with quantities of the same units in order to have the operation make sense.

Not so with multiplication and division. One can multiply or divide different units. By using these operations, one creates new units. For example, a meter is a measure of how long something is. Suppose you have a rectangle that is 3 meters in length by 2 meters in width, and you want some idea of how "big" the rectangle is. We have the intuitive idea that a rectangle that is longer than another rectangle of the same width is bigger, and similarly one that is wider than a rectangle of the same length is bigger. We also know that length itself is not a sensible measure for a rectangle. We don't say that the rectangle is 3 meters because that would ignore the width. Early mathematicians came up with the idea of multiplying the length by the width and creating a new unit as follows: 3 meters × 2 meters = 6 meter × meter, or 6 square meters, or 6 meters^2. This means that we are measuring the rectangle by using little 1-meter-by-1-meter squares as our measuring device. Sure enough, this new rectangle is made up of 6 such square meters.

Similarly, we can divide quantities that have different units in order to gain a new unit that represents a ratio. The most obvious of these is speed. Speed is distance divided by time. One can see this as follows: Someone who can cover the same distance in half the time that someone else does is clearly running twice as fast. Similarly, if he can cover twice the distance in the same time, he is going twice as fast. We measure speed in units like meters per second, which really means meters divided by seconds.

Not everything has to have units. There can be in formulas what are called pure numbers—that is, numbers without units. Suppose we have two runners, one of whom covers 10 meters in 5 seconds and the other who covers 10 meters in 10 seconds. The speed of the first runner is 10/5 meters/second = 2 meters/second. That of the second runner is 10/10 meters/second = 1 meter/second. If we call the speed of the first runner a and that of the second runner b, then $a = 2b$. In this formula a and b are both in meters/second but 2 is a pure number having no units.

Notice that the above equation—and indeed any equation—only makes sense if both sides have the same kinds of units. Just as you cannot add length to temperature, so you cannot say that a length of 6 meters is equal to a temperature of 25 degrees Celsius.

The formula for the Schwarzschild radius ($R = 2GM/c^2$) has units of length on its left side (R), so it must have units of length on its right side ($2GM/c^2$). The Schwarzschild radius by its nature of being a radius for an object of mass M whose gravitational force is going to overcome a velocity of c will depend only on the gravitational constant, mass, and light-speed values. It turns out that the only length that can be made from multiplying or dividing G, M, and c is of the form kGM/c^2, where k is a pure number. The fact that k is the number 2 in both Newtonian gravity and general relativity may be a coincidence, but the rest of the formula is not. In other words, since in both theories the only quantities needed to determine the Schwarzschild radius are G, M, and c, and since the only way to put them together in multiplication and division to get length is to do GM/c^2, we are forced to conclude that the formula will be of the kind listed above, and a little calculation returns us to the same equation for the Schwarzschild radius.

The above process may sound uncomfortably abstract for something as seemingly hands-on as gravity, but in science as in so many other parts of life what matters is to do what works whether or not it is comfortable.

Though the formula for the Schwarzschild radius is the same in both Newtonian gravity and general relativity, many other things about Newton's and Einstein's universes are not much alike. In particular, the concept of time in relativity is critically different from the usual concept of time. In everyday observation we divide up time into past, present, and future, and we think of these as matters of observation. Although in a sense, the past is detected (by the means of memory) and the future theoretical.

One can use the fact that nothing in the universe can go faster than light to understand that relativity shrinks the notions of past and future. That what is past is only what could in principle be detected in the past and what is future is only what could in principle detect the present. An occurrence is in our future not simply if it happens at a later time, but if it happens at a later time and is sufficiently close in distance that it could be reached from our present time and place by something traveling no faster than the speed of light. In other words, if light (or anything traveling slower than light) leaving us now could in any way reach an object at some time, then that object at that time is in our future, and if light (or anything traveling slower than light) leaving some object at some time could in any way reach us now, then that object at that time is in our past.

If we draw a graph of space and time (with only two dimensions of space to save the need for four-dimensional graph paper) and stick ourselves in the present at 0,0 space and 0 time, then, in Newton's view, everything above the space plane is the future, and everything below it is the past (see fig. 6a) Not so with Einstein.

The boundary of our future, those occurrences that can just barely be reached from our current time and place by something traveling at the speed of light, is called our future light cone, with the past light cone defined in the corresponding way (see fig. 6b). In the figure, these regions look like cones because we have included only two of the three dimensions of space so that we would have room for the dimension of time. Those occurrences that are neither past nor future are called spacelike separated from

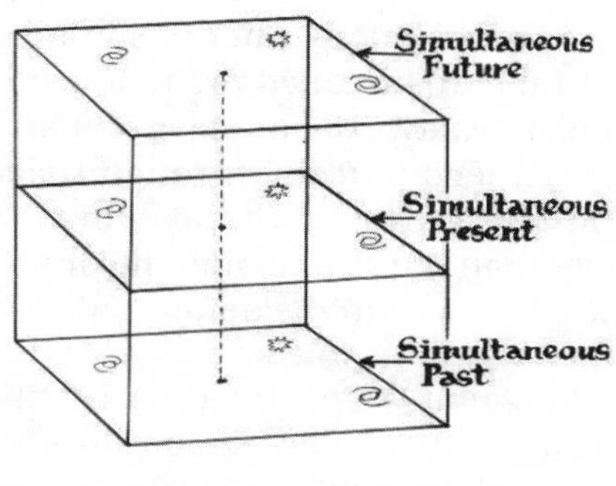

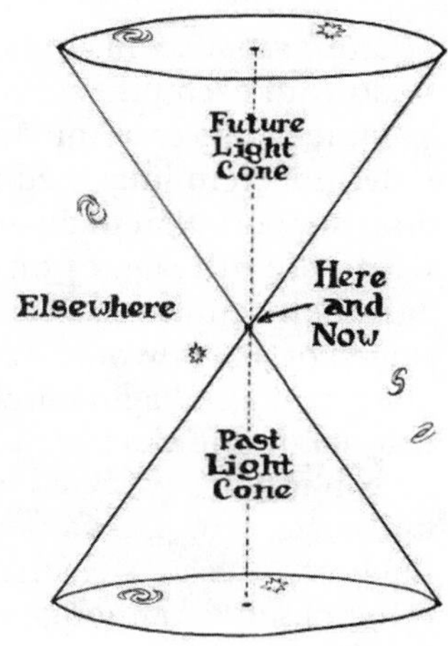

Figure 6a & b

us. For example, an event happening right now on the Moon is spacelike separated from us now, but it is in the past of us five minutes from now, since it takes a little more than a second for light to reach us from the Moon.

There are so many of these spacelike-separated events at so many different times that we don't want to call them "the present." Einstein showed that the notion of "at the same time" depends on the motion of the observer, and that for any two spacelike-separated events there is some observer who thinks that they happen at the same time. In other words, it is possible to theorize an observer whose position, speed, and direction of travel is such that any given two spacelike-separated events can seem to be happening simultaneously as far as that observer is concerned. This means that in relativity "the present" is not a useful concept, except when referring to the present of a particular observer. (Thus our remarks in the previous paragraph about an event happening "right now" on the Moon should be thought of as "right now as observed by us"). In contrast, the notions of past and future given here are independent of any observer, since what happened before and what will come after only depend on what motions are possible for objects moving slower than the speed of light.

It says something about the difference between the perceived universe and the universe of fact that something so intuitive to us as the here and now can in fact be wholly unlike our basic conceptions. This is another part of the necessity of science, because what seems to be and what is need have precious little to do with each other.

Time as a commonplace idea suffers many other hits in relativity. Einstein imagined what came to be called the twin paradox. Suppose that of a pair of identical twins, one stays on Earth while the other goes on a high-speed journey through space and then returns to Earth. Einstein showed that the twin who stayed on Earth would age more than the sibling, not only in the usual biological aging processes, but even in the amount of time elapsed on the clock.

The fact that time changes based on relative velocity is critical to the twin paradox and leads to the concept of proper time. Each observer's measurement of time is different from every other observer, although if they are close together and moving at similar speeds and directions, those times will be similar. This is the case, for example, for a large number of people standing on the same planet. They are all moving at about the same rate and experiencing about the same gravity. So despite the fact that each thing in the universe is running on different time measurements, as far as most people can tell, time is the same. Because all the observers we have ever taken records from are all (from a large perspective) in one place undergoing the same motions, they are all effectively the same observer. Six billion of us now, and not a dime's worth of difference as far as relativity is concerned.

Proper time is in effect the local time of a particular observer, the time elapsed on the watch of that observer as it follows a certain path. In general relativity, we must have a formula (called the metric) that allows us at each point in space and moment in time to determine the light cones, and that allows us for each path to compute its proper time (and thus be able to relate one person's time to another person's). Thus general relativity is a theory of space and time united into something called spacetime.

Relativity also becomes a theory of gravity with the addition of the concept that in a gravitational field an object moves along the path of maximum proper time. What this means for an observer is that high gravity also affects time and distance just as relative speed does (although with more complex calculations).

In relativity, gravity is not just a property of matter, but a shaping of the universe itself. (We will deal with this in a little more detail later on.)

This in a nutshell is Einstein's general theory of relativity. We have left out a few things, like the Einstein field equations that allow us to find the metric, and the properties of the metrics found in this way. In other words, we have left out the mathematical and theoretical apparatus needed by scientists doing research in general relativity. This brings us to one of the critical matters in learning science. Scientists need to understand their fields of knowledge in a way that allows them to expand the fields. Readers of science who seek to understand it (including scientists seeking to understand fields of science other than their own specialty) need not know the minutiae of how it works in order to be able to comprehend it. This is analogous to the difference between an auto mechanic and a driver. The driver needs to know when something is working, but need not know exactly how it is going wrong. The mechanic, on the other hand, needs to know when to throw a monkey wrench into the works. Einstein was perhaps the best wrench tosser ever. He took the car apart, stuck in some handmade components, and turned a Newtonian Stanley Steamer into a racer that clocks in at the highest speeds possible, even though it also has some of the heaviest parts, and has a cool black exterior.

Pushing that metaphor brings us back to our cool black holes.

CHAPTER 5

Black Holes in Relativity

High speed and high gravity, the conditions under which relativity becomes important, are obviously present in black holes: high speeds because not even light speed is enough to escape them and high gravity for obvious reasons. Thus though we first came to black holes in Newtonian theory, to understand them we need relativity. Michell and Laplace in their Newtonian comprehension asked what an object's properties would be if it were smaller than (what later became called) its Schwarzschild radius. They concluded that it was possible for a star to exist of a definite radius smaller that its Schwarzschild radius, but that we would not see any light from such a star because any light that it emitted would eventually fall back on the star and not escape to be seen. Such "dark stars" would be known only by their gravitational effects.

In general relativity, the light cones in the presence of an object with a strong gravitational field "tip" inward. At the Schwarzschild radius, they have tipped so far inward that all directions to the future point to a decreasing radius, and therefore all material objects move inward. Most light rays that are exactly at the Schwarzschild radius also travel inward. Only a light ray exactly at the Schwarzschild radius and pointing directly "outward" from the star does not fall in. In other words, a particle of light exactly at the Schwarzschild radius moving directly away from the center of the black hole does not escape or fall in, but everything else at the Schwarzschild radius or below falls in. The fate of that single photon is, however, rather strange. It remains at the Schwarzschild radius forever, moving as fast as anything possibly can but not getting anywhere. In *Through the Looking-Glass*, the Red Queen tells Alice that in Looking-Glass Land one must run as fast as one can in order to stay in the same place, and twice as fast if one wants to get anywhere at all. Sadly for the photon at that exact distance, running twice as fast as light is impossible no matter what side of the glass it is on.

Inside the Schwarzschild radius, everything falls toward the center, since even to stand still would require moving faster than light. In the language of general relativity, all future directions point to a decreasing radius. An object can no more escape from a black hole, nor even halt its movement toward the center, than it can go backward in time. If a star became small enough for its mass to lie inside its Schwarzschild radius, every particle of that star would continue to fall inward, so that the entire star must collapse down to a point at its center. This completely collapsed object is called a black hole, and the point is called a singularity. Gravity is so strong at a singularity that even general relativity cannot describe it; instead one would need some combination of general relativity and quantum mechanics: a quantum theory of gravity (still under construction).

Physicists call the sphere around the central point with a radius equal to the Schwarzschild radius the event horizon of the black hole. They call everything inside the event horizon the black hole interior and everything outside the event horizon the black hole exterior. Things exterior to the black hole belong to the detected universe, since signals can come to us from such objects. But what about the interior? Is it part of the detected universe, and if not, what can we do to discern that which cannot be detected?

We could only detect an event inside the black hole if some signal could reach out and get to us. But according to general relativity, such a signal would have to be moving faster than light in order to escape. General relativity says that nothing can do that, and therefore that the interior is not directly detectable as long as we remain in the exterior. (It remains to be seen whether a quantum theory of gravity also states that the interior cannot be detected.)

This is a more extreme version of the difficulty we faced with the Sun's photosphere. Light comes to us from the photosphere and therefore gives us direct information about that outer layer of the Sun, but not about the layers below it. However, in the case of the Sun, we can get indirect information from the light because the lower layers affect the light that is emitted, as well as direct information from things other than light, such as the neutrinos that come from the center of the Sun.

In the case of black holes, the prohibition of information is absolute. According to general relativity, no object and no information of any kind can pass from the interior to the exterior. As long as we are in the exterior, the interior must remain completely unknown. For this reason, the interior of a black hole is not part of the detected universe. The universe itself draws a line and says, "Beyond here you cannot see."

Black hole interiors are, however, part of the theoretical universe. To the extent that we have reason to believe that general relativity is true, we have reason to believe the answers that it provides on black hole interiors. However, as long as we stay in the exterior, we will never be able to use detections to check those answers. Furthermore, even in the theoretical universe, there is a sharp division between black hole exteriors and interiors. Since no thing and no influence can escape from the black hole, no theoretical prediction of anything going on outside the black hole depends in any way on anything going on inside the black hole. Therefore, in making theories of the exterior, we can completely ignore the interior.

If we are barred from the interior, are we then barred from applying the methods of science to black holes? Wouldn't that be annoying? Here's this really fascinating object and we can't do anything with it. Well, we can, sort of. The exterior of the black hole is still part of the detected universe and the nature of the black hole still part of the theoretical universe. This means that we can deal with the outside. We can make theories about the event horizon and the exterior. But for this to be *science*, rather than the sort of speculation that Twain complained about, black holes also have to be something we can detect. We have to draw them down from the theoretical—without, we hasten to add, actually putting them in our labs, since that would have troublesome consequences for, say, everything in our solar system.

So let's start from the beginning again. Not this time with the theory of black holes, but the theory of where they come from. Theory says that if an object were to become smaller than its Schwarzschild radius, then it would collapse to become a black hole. But do any objects actually do this? We will first consider this question for stars. Does a star ever collapse to form a black hole? We start with stars for one reason: the Sun is the biggest object with the highest gravity in our solar system, and, for reasons we will make clear later, stars can be determined to be objects like the Sun.

At first glance it seems unlikely that a star could become small enough to be a black hole. The Sun, which is a typical star, has a radius of about 7×10^5 (700,000) kilometers but a Schwarzschild radius of only about 3 kilometers. So the Sun's radius would have to be about 1/200,000th of its present radius for it to be a black hole. The situation becomes even more extreme when put in terms of density. The density of the Sun is close to the density of water. But if it shrank to its Schwarzschild radius, its density would have become over 10^{16} times larger. This is even larger than the density of the nucleus of an atom. Faced with such huge numbers, it is natural to suppose that such gravitational collapse does not happen.

But the situation changes when we flip the question around and ask not whether a star could collapse, but what keeps a star from collapsing under its own gravitational force. Recall that the Sun is in hydrostatic equilibrium; a balance exists between gravity trying to make the Sun collapse and pressure trying to make it expand. Thus the Sun staves off collapse by producing pressure.

How does the Sun make its pressure? By "burning" fuel. Hot gas (or in this case hot plasma) has pressure, and the hotter it is, the greater the pressure. From this perspective, the Sun is burning itself up trying to avoid collapse. It, too, is running as fast as it can to stay in the same place, but this method of staving off collapse cannot work forever. Eventually a star must use up all its fuel, fusing away the ability to maintain pressure. Yet though fuel goes, gravity remains, since the residue of the burned-up fuel is still part of the Sun. In effect, the Sun's fusion is racing forever against an opponent that does not tire, gravity, inexorable and indifferent to needs for power. Eventually in such a race, the racer that can tire will become exhausted and the unwearying opponent will win.

It would seem then that a star would inevitably form a black hole. Following this logic, every object that has no source of power would become a black hole. Earth would become a black hole. But Earth does not because it has other, non-fuel-based pressures, such as the solidity in the crust and the hydrostatic pressure of the mantle, to keep it from collapsing. We are thus led to ask the question "What source of pressure, if any, does a star still have when its fuel is used up?" In other words, is there something that will still stave off collapse even if the pressure of burning is gone?

At first we might guess that the ordinary chemical forces between atoms, as in a solid or a liquid, might do the trick. These forces tend to be attractive, pulling atoms together until the atoms reach a certain density and then resisting further compression as the atoms repel each other. We know that electromagnetism is stronger than gravity, so we might think that here is a candidate to keep the pressure on even after the fuel is gone. Thus we might guess that a star when it runs out of fuel simply becomes a solid or liquid, and that the ordinary pressure of such substances is sufficient to balance gravity. This is what happens for Earth, which is in hydrostatic equilibrium but does not burn any fuel to generate its pressure. However, the Sun is much more massive than Earth, and gravity in the Sun is much stronger. The puny chemical forces that serve to balance Earth's gravity are much too weak to do the same for gravity generated by the mass of the Sun. The Sun's material remains a plasma, even when compressed to densities beyond that of ordinary solids and liquids.

What, if any, other source of pressure is available to a star? There is something called degeneracy pressure, which has to do with two properties of quantum mechanics: the uncertainty principle and the Pauli exclusion principle. The uncertainty principle formulated by Werner Heisenberg (1901–1976) is the bugaboo of quantum mechanics and the primary source of misunderstanding of this subtle field of physics. It has also been exploited in the creation of several pseudosciences that are clothed in a veneer of quantum theory. The uncertainty principle as usually stated says that the more accurate you are in measuring the position of a particle, the less accurate you can be about measuring its velocity and vice versa. In other words, there is a fundamental limit on the precision with which the position and speed of an object can be measured. The whys and wherefores of the uncertainty principle would—ironically—take up too much time and space for us to deal with here, but suffice it to say that while it puts a limit on observation, it can itself be experimentally verified. In other words, uncertainty is not just a part of the theoretical universe; in the process of defying detection, it itself can be detected.

For our purposes, we are going to restate the uncertainty principle as saying that on the small scale objects are always in motion. The smaller the space in which the objects are confined, the faster they move. If you trap a particle in a really small box, it will be flying all over the place in that box. Since confined fast particles push against anything that comes in contact with them, they create a pressure. So on a small-enough scale, uncertainty alone creates a pressure.

The Pauli exclusion principle has to do with the energy levels, or quantum states, for electrons (treated in the last step in the context of atoms). As we said before, electrons, whether in atoms or not, can have only certain quantum states. The Pauli exclusion principle says that only one electron can occupy each state. If all states of a given energy are filled, then the next electron must occupy a state of higher energy.

Together these two principles say that if we take a certain number of electrons and try to confine them to a smaller space, we will have to supply energy to them to account for the higher energy levels they are forced to occupy. This means that a gas of electrons, no matter what its temperature, resists being compressed because you must put in more and more energy to keep compressing it (since the electrons are filling the lowest allowed states, and by the uncertainty principle, the smaller the space the electrons are confined to, the larger the speed and therefore the larger the energy of each allowed state). In other words, a gas of electrons has a pressure, even when it is not hot. This pressure is called degeneracy pressure.

You might think that this pressure has to do with the fact that electrons repel each other electrically. However, that is not the case. The electrons in stars are in a plasma with an equal number of protons, so the total electric charge is zero. The repulsion between electrons and electrons (and between protons and protons) is completely canceled out by the attraction between electrons and protons. Yet there is still a degeneracy pressure. Furthermore, even neutrons, which have no charge, have a degeneracy pressure, since they, too, are governed by the uncertainty principle and the exclusion principle.

A plasma has protons and electrons, so we should expect a star to have proton degeneracy pressure too. However, the proton is almost 2,000 times as massive as the electron, and all other things being equal, the larger the mass of the particle, the smaller the degeneracy pressure. Stars do have both electron degeneracy pressure and proton degeneracy pressure, but the proton degeneracy pressure is negligible.

A star that is no longer undergoing fusion but is kept from collapse by electron degeneracy pressure is called a white dwarf. A star must be compressed to a very high density for electron degeneracy pressure to become strong enough to counteract gravity, but this high density is still nowhere near that needed for a black hole. A typical white dwarf has about the mass of the Sun and is about the size of Earth.

We might guess that all stars when they use up their fuel become white dwarfs. However, in the 1930s the astrophysicist Subrahmanyan Chandrasekhar showed that this was not the case. Though born in India and educated in England, Chandrasekhar spent most of his career in the United States. He made many contributions to theoretical astrophysics, but his study of white dwarfs is what he is best known for. With an encyclopedic grasp of physics and no fear of even the most difficult calculations, Chandra (as he was known to his colleagues) worked on fundamental problems in astrophysics and general relativity until his death in 1995 at the age of eighty-four. His book *The Mathematical Theory of Black Holes* has some formulas that are longer than a page. When doing calculations, he often held the paper sideways because the normal width of paper was not wide enough for the formulas he wanted to write.

Chandrasekhar studied and considered the contest between gravity and pressure. Suppose, as he did, that at some time a star has a degeneracy pressure that is too small to balance gravity. Then the star will contract under the force of gravity and the degeneracy pressure will get stronger (smaller box, faster motion, more energy). Eventually, one might think, the star will contract enough that pressure and gravity balance. However, since the star has contracted, its parts are closer to each other than they were before. By Newton's formula for the gravitational force, this means that the gravitational force compressing the star is stronger.

In other words, as the star contracts, the opposing forces, gravity and pressure, each get stronger.

It is not clear whether at any point in this process these forces will come into balance. To answer this question, Chandrasekhar did a detailed calculation using the equation of hydrostatic equilibrium and the equation for the degeneracy pressure of an electron gas. The result of this calculation is that there is a maximum mass for a white dwarf star. This maximum mass, now known as the Chandrasekhar limit, is about 1.4 solar masses. If the mass is larger than the Chandrasekhar limit, the pressure is never strong enough to resist gravity. A white dwarf can have a mass below the Chandrasekhar limit, but no white dwarf can have a mass above the Chandrasekhar limit.

We might now guess that there are two possible end states for stars. A star that starts out with a mass below the Chandrasekhar limit will end up as a white dwarf, while one whose mass starts out above the Chandrasekhar limit will become a black hole. However, this guess is not quite right because of three complications: one an everyday solar phenomenon, stellar mass loss; one an incredibly dramatic stellar event, supernova explosions; and one a quantum process that we mentioned in passing: neutron degeneracy pressure.

The first of these, stellar mass loss, is a simple, detectable phenomenon. Recall that the Sun is ejecting mass from its outer layers as the solar wind. For the Sun as it is now, this mass loss is small enough to be negligible. However, stars in the later stages of their lives eject mass at a much faster rate, so a star that begins with a mass above the Chandrasekhar limit can eject enough matter to get below the Chandrasekhar limit and end as a white dwarf.

Supernovae (the plural of supernova) are stellar events that while rare have been recorded and noted off and on over the centuries. A supernova can light up the night sky for a time, a prima donna upstaging all the other stars by producing its own spotlight. To understand a supernova, we need to examine in a little more detail what happens to a star that is too massive to be a white dwarf when it has used up its fuel. It is helpful under these circumstances to think of the star as composed of two pieces: A very dense "core," which is the central region of the star, and a less dense "envelope" consisting of the rest of the star. This is a simpler solar model than the ones we used previously, but for these purposes it works quite well. This is also a principle of science: if you don't need the more complex model, don't use it. Newton's laws of motion work fine most of the time—we don't need differential geometry to calculate how fast a car is going. Similarly, we don't need to know exactly all the layers of a star to determine what happens when it explodes.

In a star that has exhausted most of its fuel, the core of nuclei and degenerate electrons collapses when it reaches the Chandrasekhar limit. In the process of the collapse, the electrons combine with the protons to make neutrons and neutrinos. What is left is a core of neutrons and a very large release of energy. This energy is originally carried by the neutrinos created in the above process, but many of these neutrinos are absorbed, producing forms of energy that interact rather more often than those neutrinos, so instead of the blithe transmission of indifferent particles, the energy is transmitted to the star's envelope in a gigantic explosion called a supernova. Much of the envelope is ejected in this explosion, and the rest falls back to join the core. Thus a supernova is two processes in one: the gravitational collapse of the core and the explosive ejection of the envelope. It is the energy released by the collapse of the core that powers the explosion. The core is then a very dense collection of neutrons at the density of atomic nuclei. It is an object that is generally more massive than the Sun, but smaller around than a good-sized city. This object is called a neutron star.

What keeps a neutron star from collapsing under its own enormous gravity? Neutron degeneracy pressure. We passed over it before because when large numbers of electrons are present (as they are in white dwarfs), electron degeneracy pressure is much stronger than neutron degeneracy pressure. Neutron stars are made up of neutrons, and there's only the neutron degeneracy pressure keeping them from falling farther. However, just as there is a maximum mass for a star held up by electron degeneracy pressure, so there is a maximum mass for a star held up by neutron degeneracy pressure.

The calculation of the maximum neutron star mass is more involved than for white dwarfs, because one must also take into account the strong force. That is, it is not completely accurate to think of a neutron star as a gas of neutrons. One might think of it as a gigantic nucleus held together by gravity, but with the strong force still important for the interactions between each neutron and nearby neutrons. The details of these interactions at neutron star densities are not completely known, in part because doing calculations with the theory of the strong force runs into complications that we don't want to dig into now. However, what is known about the strong force, both through the theory of the strong force and experimental tests of nuclear reactions, is enough to make estimates of the maximum neutron star mass. These may need revision at some time when there is greater understanding, but they are likely to be roughly accurate. These calculations give an estimated maximum neutron star mass of about 2 solar masses. But remember, this is not the initial mass of the star, but the remaining mass after stellar mass loss and the ejection of a lot of the envelope in the supernova.

We are now in a position to give an at least theoretical answer to the question of whether or not black holes can be formed by stars. Consider a sufficiently massive star that undergoes a supernova explosion where the initial neutron core plus that part of the envelope that is not ejected are more massive than the maximum neutron star mass. Such a star undergoes complete gravitational collapse and becomes a black hole. Turning this around, we have a theoretical explanation of where black holes come from. If a star has sufficient mass and retains sufficient mass after collapse and becoming a supernova, then (in theory) it will become a black hole.

Thus we have three possible end states for a star: it becomes a white dwarf, a neutron star, or a black hole. Which of these occur depends on the initial mass of the star, with the lightest stars becoming white dwarfs, those of intermediate mass becoming neutron stars, and the very heaviest becoming black holes.

It might be considered unfair to talk only about the deaths of stars (particularly since our lives depend so much on the life of one in particular), so in the interests of fairness, charity, and avoiding an overly morbid view of the universe, let's look at their whole lives, shall we?

CHAPTER 6

The Lives of Stars

A star begins as a cloud that consists mostly of hydrogen and helium gases, which under the force of its own gravity collapses inward. As the gases come together, the cloud becomes hotter through the Kelvin-Helmholtz mechanism. Eventually the atoms in the cloud become hot enough (hence fast enough) that nuclear fusion of hydrogen begins, a process that adds to the heat of the cloud, sparking even more fusion, until so much of the cloud is fusing it becomes a star. During this stage of a star's life, as we noted, it puts out energy at a decent regular rate, so that if there happened to be, say, any planets around it, they would be bathed in a continuous flow of solar energy that might be of some use to meteorological and chemical processes on the planet. Some of these processes might indeed make highly sophisticated uses of this solar energy, producing perhaps climates and life and other stuff like that. But for the life cycle of the star, those are mere side effects, not concerns.

The star spends most of its lifetime in this hydrogen-burning stage. The end product of hydrogen fusion is helium (along with the photons and neutrinos we already mentioned). The helium produced accumulates in the core of the star. This helium is providing mass that contributes to the gravitational attraction, but is not itself undergoing fusion and so is not providing thermal pressure to help withstand the gravitational attraction.

Eventually enough helium accumulates so that pressure becomes weaker than gravity and the helium core collapses. It is helpful to think of this helium core collapse as a repeat of the Kelvin-Helmholtz mechanism that led to the formation of the star. As the core collapses, it becomes hotter. This heat produces pressure that helps to stave off (or at least slow down) the collapse process.

Eventually the core becomes hot enough for the helium itself to begin fusing. Helium fusion requires a hotter temperature than hydrogen fusion for two reasons. First, each helium nucleus has twice as many protons as a hydrogen nucleus, so two helium nuclei repel each other with four times the force of two hydrogen nuclei, therefore the helium nuclei need to be more energetic in their movements in order for them to collide rather than push each other away. The other reason is that helium fusion is a more complex process involving the combination of three helium nuclei to make a nucleus of carbon. Why doesn't helium fusion simply combine two helium nuclei? When two helium nuclei combine, they form an isotope of beryllium with four protons and four neutrons. However, this isotope of beryllium is unstable (for reasons of nuclear physics that we are not going to go into here) and quickly breaks up again into two helium nuclei.

The fusing of helium to form carbon can be viewed as a two-step process. First, two helium nuclei combine to form beryllium. If before the beryllium can break up again it combines with a third helium nucleus, carbon will form and this carbon will be stable. Once carbon has formed, it can fuse with yet another nucleus of helium to form stable oxygen. Thus during the helium fusion stage, the core of the star becomes increasingly rich in carbon and oxygen. In contrast to beryllium, carbon and oxygen are stable nuclei. When the star is burning helium, it is producing energy at a greater rate than when it was burning hydrogen. That is, the star is emitting more light. This greater power swells the size of the envelope and the star becomes what is called a red giant. This tends to be bad news for any planets that had been happily enjoying the sedate shower of hydrogen-fusion photons.

What happens next depends on the initial mass of the star (that is, the mass the star had when it first started fusing hydrogen). For stars with an initial mass less than about 8 solar masses, the core never becomes hot enough to fuse carbon. Instead, the envelope is gradually ejected and the carbon-oxygen core settles down to become a white dwarf, which goes on to a quiet, crotchety old age.

However, a star with an initial mass greater than 8 solar masses has greater pressure and therefore greater temperatures, so the carbon-oxygen core eventually becomes hot enough that carbon can fuse with carbon and oxygen with oxygen. These reactions produce a host of heavier elements, including neon, sodium, magnesium, silicon, phosphorous, and sulfur.

One might think that this sort of reaction could go on and on generating energy by producing ever-heavier elements. But recall the analogy between the Sun and a hydrogen bomb, and then recall that there is another type of nuclear bomb (often called an atomic bomb) in which the energy is provided by a heavy nucleus (uranium or plutonium) splitting apart into two lighter nuclei. These two bombs form a very strange contrast. How can energy be generated both by fusion (the combining of two or more nuclei into one) and fission (the splitting of one nucleus into two or more nuclei)?

A nucleus can be thought of as having a binding energy: the energy that it would take to break the nucleus apart. A nuclear reaction gives us energy if it takes nuclei with a certain binding energy and rearranges them into nuclei with even more binding energy. A nucleus has binding energy because of the strong force holding the protons and neutrons together, but that binding energy is lessened by the electromagnetic force that pushes the protons apart. The binding energy is also lessened by the fact that each proton and neutron needs to be in a certain energy state and that as lower states fill up, more energy is needed to put one in a higher state. What this adds up to is that the lightest and the heaviest nuclei are the most weakly bound, and it is nuclei of an intermediate mass that are the most tightly bound. Thus fusion of light nuclei and fission of heavy nuclei both yield energy.

Among all the elements, there is a particular nucleus with the most binding energy: iron. Iron takes energy to fuse or to fission. Atomically, iron is an energy sink, completely useless (chemically, of course, we red-blooded creatures can't live without it). So in the process of stellar evolution fusion, the engine of stars can give energy only when producing elements lighter than, or as light as, iron. Once a star produces iron in its core, that iron cannot be used as fuel. This does not mean that elements heavier than iron cannot be made, but simply that it takes even more energy to make them. Iron accumulates until the core reaches the Chandrasekhar limit and then the core collapses, producing a supernova explosion.

This explosion is actually the factory that produces all the elements heavier than iron. The force of the supernova slams neutrons into the heavy nuclei to produce even heavier nuclei, and then some of the neutrons in these heavier nuclei undergo beta decay, yielding protons. Remember that an element is defined by the number of protons in its nucleus. If you slam one neutron into iron (atomic number 26) and then that neutron decays, the iron becomes cobalt (atomic number 27). Slam in another and it will become nickel (atomic number 28), and so on. (Actually it's messier than that, but you get the idea.) Notice that this is happening in the star's envelope where the force of the supernova is expressed, so as the matter is flying away from the supernova, it is transmuting into heavier elements.

After collapse, the core becomes either a neutron star or a black hole, depending on how much mass from the envelope falls back on it. It is estimated that a star with an initial (at the time of first hydrogen burning) mass of between 8 and about 25 solar masses will eventually become a neutron star, while a star with an initial mass greater than about 25 solar masses will eventually become a black hole.

This history of stars is also the story of the origin of chemical elements. All elements heavier than helium are produced in stars, and all elements heavier than iron are produced in supernova explosions. In particular, every atom in our bodies except for the hydrogen atoms has been produced in the core of a star.

How do we know this is the origin of the elements? We can to some extent check. Our own solar system is made up of material processed in previous generations of stars. Solar spectra as well as analysis of the chemical composition of meteorites give us information about the relative abundances of elements (such as how many atoms of carbon there are for each atom of iron) in the solar system. Using information from the light that stars give off, we can also make "stellar models" for stars in analogy to the "solar model" for the Sun treated in the last chapter. These stellar models include nuclear fusion and yield predictions for the relative abundances of elements. These predictions are in good agreement with the data from the Sun and meteorites. Though indirect, the model holds together under tests from multiple directions, so we have good reason to believe what it says.

This last point is important because some parts of science deal with things that happen over times much longer than human lifetimes. The narratives of the results of such science often have the feel of Rudyard Kipling's *Just So Stories,* where the story starts with a state unlike things as they now are (such as trunkless elephants) and ends up with the state we know today (huge noses!). The reader is likely to have similar reactions to both Kipling and this science, something to the effect of "That's a very amusing story, but of course it was long ago and far away, and you weren't there. There is no reason to believe that the story is true."

But unlike the *Just So Stories,* the science to explain something long ago and far away is derived from things that can be tested here and now. Under the uniformity of nature, what holds now held then. Fusion in the H-bomb and in a reactor are the same processes as fusion in a star. So what we measure in the lab and in the sky now we can use to model what had been happening then. The science is the result of careful quantitative modeling that is only accepted if it matches the result of careful quantitative detection. The reporting of the end result may sound like Kipling, but these stories really are true. Besides as we will see later, we have a means of looking back in time, and what was long ago can be seen here and now.

✸

So now we have a map of stellar life that asserts the conditions under which a star can become a black hole. But in order to reach that goal, we relied on a set of assertions about the properties of stars without properly justifying those assertions. We said that stars are suns and that what we learned about the ways of the Sun is true of them as well. But we now return to the question of the last chapter and indeed of this entire book. How do we know that any of this is true and not merely a "just so" story?

Why do we today think that stars are objects like the Sun? For most of human prehistory and history, the assumption was that stars were small lights in the sky and the Sun was a big one, and more importantly that they were fundamentally different kinds of objects. Why would anyone think they were the same? Just look at them; apart from light, what do they have in common? It's not as if we get heat from stars, just twinkly light. How can they be like the Sun?

More broadly, is there some circumstance under which things that look very different are actually the same? In the perceived universe, we often notice that a seemingly small object is really a big object at a distance. If the Sun were at a much larger distance from us, it would look like a star. This gives us an initial hypothesis: Stars are far-off suns. To determine whether stars are really objects like the Sun but at much larger distances, we need to know how far away stars are.

Recall that the distance to the Sun was measured using the technique of parallax. An object seen against a distant background has two different angular positions when seen from two different places. One-half of the distance between the two points of observation is called the baseline, and one-half of the difference of the angular positions is called the parallax. The parallax is the ratio of the baseline to the distance that we want to measure (hence the distance is equal to the baseline divided by the parallax). In order to effectively and accurately measure an astronomical distance, we either need to establish a large baseline or be able to accurately measure a very small angle or, preferably, do both.

We usually measure angles in degrees, with a full circle, the largest angle possible, having 360 degrees. But a degree is much too large an angle for the parallaxes of stars, so astronomers introduced the units of arc minutes and arc seconds. There are 60 arc minutes in a degree and 60 arc seconds in an arc minute. So an arc second is 1/3,600th of a degree, a very small angle. Arc seconds are convenient units for parallax. Existing telescopes can be used to accurately measure angles down to the arc second, so we are okay on that front. Note that arc minutes and arc seconds have nothing to do with time; the reason for the name is simply that they divide up a degree in the same way that ordinary minutes and seconds divide up an hour.

What about the baseline? To measure the AU, Richer and Cassini used two different locations on Earth, but trying to measure stars this way gives a parallax too tiny for us to measure. This tells us immediately that stars are very far away, which gives a hint that perhaps they might be sunlike. But we need to do more than say that they're a long way off. We need accurate measurements.

The trick that solved this problem was to use the AU itself as a baseline. Earth goes around the Sun in one year; if we make two observations six months apart, then Earth at the time of the second observation is 2 AU away from Earth at the time of the first observation. In other words, the baseline of this measurement is 1 AU. This is a very efficient experimental apparatus. All that is needed in order to travel the needed distance is to wait between measurements.

If we take two photographs of the same field of stars, one picture in January and one in July (for example), and we put the two photographic plates on top of each other, the parallax image is created. When we do this, most of the stars (the distant ones) line up with each other; but a few stars (the nearby ones) have slightly different positions from one photograph to the other. From this tiny difference in position on the photograph, and from the known angular size of the star field that we photographed, we find the parallax of the star. For a baseline of 1 AU, a star with a parallax of 1 arc second has a distance of about 206,000 AU. Since an AU is about 93,000,000 miles, this means that a star with a parallax of 1 second is about 19,200,000,000,000 miles away from us.

This is one of those very large astronomical numbers that we will make disappear by the simple expedient of creating a new unit of length. Just as we made 93,000,000 miles vanish into 1 AU, we will prestidigitate 19,200,000,000,000 miles into a new unit called a parsec, which is defined as the distance needed to produce a parallax of 1 arc second using the method given above. This terminology is a bit confusing, since stars at larger distances have smaller parallaxes. A star with a parallax of 1/2 second has a distance of 2 parsecs, 1/3 second 3 parsecs, and so on. A parsec is a convenient unit of distance for stars, since the nearest stars are within a few parsecs of us, but there are also other convenient units for measuring such distances and indeed greater distances. We will use the speed of light (a universally convenient quantity) to create distance measurements.

Light travels 186,000 miles/second or 300,000 kilometers/second. We define 1 light-second as the distance light would travel in 1 second, 186,000 miles. One light-minute is the distance light would travel in 1 minute, which is 60 light-seconds or 11,160,000 miles or 18,000,000 kilometers. By analogy, we can create the light-hour, light-day, light-week, light-year, light-century, and so on. In astronomy the light-year is often used; the others generally are not. Light takes about 8 minutes to get from the Sun to us, traveling a distance of 1 AU, so 1 AU is roughly 8 light-minutes. Since a parsec is roughly about 200,000 AU, this means that it takes light about 1,600,000 minutes or about 3 years to travel a distance of 1 parsec. A more accurate value is 3.26 years, so a parsec is about 3.26 light-years.

The nearest star to us (apart from the Sun, you smart alecks) is at a distance of about 1.3 parsecs, or about 4.3 light-years. It is helpful to consider these units when thinking about how long it would take to travel to the stars. The space shuttle orbits Earth in about 90 minutes, but light can travel around the Earth about 7 times in 1 second. So the space shuttle is traveling only about 1/40,000th the speed of light. It would take a spaceship traveling at the speed of the space shuttle about 170,000 years to get to the nearest star. This presents a few logistical problems for interstellar travel. This is also why science fiction writers usually cheat and invent physically impossible means of going from star to star.

As a side note, the only unfortunate characteristic of the light-second/light-year terminology is that it sounds to our ears like units of time rather than distance. We hear "light-year" and we think "year." This has caused a certain number of silly lines of dialogue to appear in some science fiction movies and TV shows that shall remain nameless. Oddly enough, as will become clear later in this chapter, there is a use to be made of the mental confusion this creates. Since we see things by means of light, an object that is 8 light-minutes away (to pick a big, shiny object at random) is showing us images of itself from 8 minutes ago. In other words, the distance measured in light units gives us how far into the past we are looking when we see it. This will end up being important later when we discuss quasars.

The largest distance that we can measure using the parallax method depends on the smallest angle that we can measure. For a telescope, the larger the diameter of the telescope mirror, the smaller the angle that can be measured. The distortion of Earth's atmosphere also places a limit on the smallness of angles that can be measured from the ground. From the ground, the smallest angle that can be measured with any accuracy is about 0.01 arc seconds, yielding parallax measurements of up to about 100 parsecs. The smallest angles must be measured from space. In space, the Hipparcos satellite has accurately measured angles as small as about 0.001 arc seconds, corresponding to distances of up to about 1,000 parsecs. This is a long cab ride, but nowhere near long enough for many of the objects we want to study. In the next step, we will consider how distances larger than this can be measured. For now, parallax gives us all we need.

The parallax method tells us that even the closest stars are so far away that they must be either big or bright or both. This makes the idea that they are sunlike plausible, but we need more information to be sure. In particular, we need to know how bright stars are.

A brief digression: We have been talking about the distance to the stars as if it were one number, but we now know that each star is at a different distance from us than the other stars. This in itself was a radical discovery, because if you look at the night sky, the stars seem to be the same distance away. The ancient Greeks had an image of stars as little points of light on a huge sphere centered around Earth. To them, all stars were the same distance from Earth. We now have detected that this is not the case. Each star has its own placement, an object moving in the sky without anything holding it near to Earth. As we go on, we will find that each star will also have its own luminosity and other characteristics. We quickly discover that stars are much more individual than they appear at first; but still those individualities belong to a broader commonality of stellar life.

Once the distance to a star is known, its luminosity can be found using a similar method to that used for the Sun. Pointing a telescope at the star, we measure the light collected by the telescope. Then, using the diameter of the telescope mirror and the distance to the star, we calculate what fraction of the total light put out by the star reaches the telescope and thus find the luminosity of the star.

The actual measurement of stellar luminosity works the same way a digital camera does, minus the last "user-friendly" step. In a digital camera, light is focused by the lens on something called a CCD (charge-coupled device). The CCD consists of many light-sensitive elements called pixels (a 5 megapixel camera has 5 million pixels). In each pixel, electric charge builds up proportional to the amount of light received by the pixel. The amount of electric charge from each pixel is then read off and stored as a number.

In a digital camera, the last step is to reverse everything and make an image from this set of stored numbers. In the measurement of stellar luminosity, the numbers are instead kept as numbers. The numbers from all the pixels receiving light from a star are added up and then used with the diameter of the telescope mirror and the distance to the star to calculate the luminosity of the star. In this way it was found that the star Sirius A has about 23 solar luminosities (it is putting out about 23 times as much power as the Sun), that Alpha Centauri A has about 1.5 solar luminosities, and that Tau Ceti has about 0.6 solar luminosities.

These luminosities are on the same scale as that of the Sun (by the same scale, we mean that they aren't hugely larger or smaller), so now we know that stars are sunlike in more characteristics and our hypothesis that they are suns gains more plausibility. If we can measure their size and temperature and come again near to those of the Sun, we will be strongly justified in our hypothesis.

We found the Sun's size from its angular size (how large it looks) and then found the temperature from its size and luminosity. However, this won't work for stars. They are so far away that we cannot measure their angular size. Even in our most powerful telescopes, stars are still just points of light. Instead, we reverse this process to first find the temperature of a star and then use the temperature and luminosity to calculate its size.

A star's temperature can be estimated from its color. As we noted above, the hotter an object is, the shorter the average wavelength of light it puts out, and that for light, wavelength is the same as color: cool stars are red; hot stars are blue. However, a more accurate method comes from a star's spectral lines. Remember that a spectral line comes about when light causes an atom to change from one energy level to another. The strength (intensity of the color when in the spectrum) of that spectral line depends on how many atoms were in the first energy level. The number of atoms in a particular energy level in turn depends on temperature. At low temperatures, an atom is likely to be in its lowest possible energy state; at higher temperatures, the chance of an atom being in a higher energy state increases; and finally, at even higher temperatures, an atom is likely to lose one or more of its electrons. Spectral analysis of stars and the interstellar medium (the gas and dust between stars) shows that the chemical composition of a star is about three-quarters hydrogen, one-quarter helium, and about 1 percent everything else. Notice that this also helps confirm the "stars are suns" hypothesis since the above fits with our Sun's composition.

One might expect that the strongest spectral lines of a star would be hydrogen lines. It seems surprising that in the Sun, calcium lines are stronger than hydrogen lines. But hydrogen in its lowest energy state does not absorb light in the visible band; and in the Sun's photosphere, most of the hydrogen atoms are in their lowest energy state. To have prominent hydrogen lines, a star must be hot enough that a hydrogen atom is likely to be in higher energy states, but not so hot that it is likely to lose its electron.

Similar considerations apply to the lines of other elements. A careful examination of the strengths of the spectral lines of a star allows astronomers to find the temperature of the star (by answering the question "How hot would it have to be for these intensities to appear?"). The stellar spectra are classified as type O, B, A, F, G, K, and M (which generations of astronomy students have memorized using the somewhat sexist mnemonic "Oh be a fine girl, kiss me"). Here O is the hottest and M is the coldest. The Sun is a G-type star.

This sort of terminology is an example of the confusion that can arise from using the terms of the detected universe. If you were going to devise a way to classify stars according to their temperature, you might label stars A, B, C, D, E, F, and G, with A the hottest and G the coldest; but you wouldn't chose O, B, A, F, G, K, M. Why did the astronomers who made up this scheme?

Regrettably, they had accumulated many stellar spectra before they knew what the data meant in terms of stellar temperatures. Remember spectra are fairly easy to observe, but correct interpretation of easily observed phenomena often comes much later when a theory that fits the facts is created. It was natural to classify the spectra in terms of the strength of certain spectral lines, especially the hydrogen lines, with A being the type with the strongest hydrogen lines, then B, and so on. Later, when the connection between spectrum and temperature was learned, it was understood that for a classification of temperature many of the distinctions of the previous classification were not needed (that is, that categories could be merged and some of the corresponding letters discarded) and that the order of the letters needed to be changed. We were left with O, B, A, F, G, K, M. This kind of problem arises when terms are created for one use and shifted over to another. Unfortunately, because every astronomer had to learn this system, there has never been a concerted effort to reform it. This is one of the ways in which the artifacts of earlier ways of thinking can remain even when the ways themselves have been replaced.

We now know the distance, luminosity, temperature, and therefore the size of stars, all of which fit the "stars are like the Sun" hypothesis. What would clinch it would be to know the mass of stars. This is based on a fundamental view of science that two objects with the same characteristics are the same kinds of object. Recall that for the Sun the mass was measured by using the orbits of the planets and the formula $M = rv^2/G$, where M is the mass of the Sun, r is the radius of the planet's orbit, and v is the speed of the planet. We might want to try to apply this to a star, using the planets in orbit around that star. Unfortunately, planets outside our own solar system are very hard to detect. Indeed, this has only recently been done successfully, first by the astronomers Geoff Marcy and Paul Butler, and even then the planet they detected was found indirectly by its gravitational effects on the star, which is what we were going to use to find the mass of the star. We need another method.

To accomplish this, astronomers make use of the fact that many stars are really binary star systems, two stars so close together as to form one system. In the solar system, we usually say that the Sun does not move and the planets travel in orbit around it. However, just as the Sun tugs on the planets, so the planets pull on the Sun. In the solar system, the Sun moves in a small orbit of its own due to these pulls. It is exactly this sort of small orbit, in the case of a star, that the planet hunters, beginning with Marcy and Butler, detect when they find planets in orbit around other stars.

In a binary star system, there are two stars, each of which exerts a gravitational pull on the other. These pulls cause each star to travel in an orbit. We might say that in a binary star system, each star is in orbit around the other. It is not easy to spot which star systems are binaries, because just as our solar system is several AU across, so is a typical binary system. The angular size for a binary system (how large it appears to us) would then be something like an AU divided by a parsec—in other words, a very small angle that we can't resolve with the naked eye and often have trouble resolving even with a powerful telescope. Some nearby binary star systems can be resolved in the telescope as two separate stars. For others, it is noticed that the system is a binary because its spectrum is a combination of two different stellar spectra. For still others, the binary nature of the system becomes clear as one star passes behind the other (an eclipse) and the light coming from the system goes down until the star emerges. Finally, the orbital motion of the stars due to their gravitational effects on each other can be measured.

For a binary star system, a slightly more complicated version of the mass formula given above allows us to find the masses of both stars in a binary system given the speeds and sizes of the orbits of each of the two stars. In the case of the Sun, we found the speeds of the planets by using the size of the orbit and the period (the time it takes the planet to go around the Sun). For some binary star systems, this method will also work, but for others it is better to measure the speed and the period and use these to calculate the size of the orbit.

But how to measure the speed of a star? The method is essentially the same one that the police use to measure the speed of a car: the Doppler effect. You can see—or in the following example hear—the Doppler effect the next time a fire engine with its siren on passes by you. Listen carefully to the siren. You will notice that the siren is more high-pitched when the fire engine is coming toward you and more low-pitched when the fire engine is moving away. What is happening is that the wavelength of the sound waves that an object emits changes when the object moves. The wavelength becomes shorter when the object is moving toward you and longer when the object is moving away. The faster the object is moving, the greater the change. Thus by measuring the change in wavelength, you can find the speed. This is exactly what the police do to catch speeders, though in this case with radar rather than sound. The radar gun fires a pulse of radio waves that bounce off the moving car and come back to the radar gun. Inside the radar gun, the difference between the wavelengths of emitted and reflected waves is found and the speed is calculated.

For stars, the speed is calculated using the Doppler shift of spectral lines. We already know the wavelengths of the needed spectral lines as these have been measured in the lab. When measuring the same spectral line in a star, we see that it has a slightly different wavelength than that of the elements in the lab. This difference is called a redshift if the wavelength from the star is longer (pushed toward the red side of the visible spectrum) and a blueshift if the wavelength is shorter (pushed toward the blue end of the visible spectrum). A redshift comes from a star moving away from us, and a blueshift comes from a star moving toward us. From the size of the redshift or blueshift, we can find the speed of the star. A star in orbit will sometimes be moving toward us and sometimes away from us. Looking at a spectral line of the star over time, we will see the shift in wavelength go up and down from red to blue and back to red. The period of this up-and-down motion is the period of the star's orbit, while the magnitude of the shift in wavelength allows us to find the speed of the star.

If we put together this spectral speed data with the formula given above, we have enough information to find the mass of the star, and we find out that again they are within the order of magnitude of the Sun (an order of magnitude is a factor of 10, so 90 and 10 are within one order of magnitude of each other). What we are saying is that stars are mostly no more than 10 times more massive than the Sun and mostly no less than 1/10th the mass of the Sun. Having accumulated so many pieces of data in common, we can now be confident that stars are suns and that therefore our discussion of black holes, neutron stars, and white dwarfs is more than theory and can well be applied to the universe itself. We have now made sure that our second step is on firm ground.

✷

Now that we know the luminosities, temperatures, and masses of many stars, what can be done with the information? Having established how much they are alike, we might try to figure out what the differences between the stars are. One way to start this process is to make what is called a scatter plot of the stars' luminosity and temperature, thus giving a shape to their variations. This scatter plot is a graph with temperature on the horizontal axis and luminosity on the vertical axis. Since each star has a temperature and a luminosity, each star can be represented by a single point on the graph. All the stars for which temperature and luminosity are known then become a set of points on the graph. This sort of plot is called an H-R diagram after the astronomers Ejnar Hertzsprung and Henry Russell, who first made such a scatter plot of temperature and luminosity in 1914. Hertzsprung was a Danish astronomer, best known for finding the relation between color and brightness of stars that forms the basis for the diagram that bears his name. He also found the luminosity of Cepheid variable stars, which are important for finding the distance to galaxies. Hertzsprung had no formal training in astronomy but was instead a chemical engineer studying the chemistry of photography. Applying photography to the measurement of starlight gave the result that he is known for. Russell was an American astronomer, best known for finding the relation between luminosity and spectral type that forms the basis for the diagram named after Hertzsprung and him. He also studied binary star systems and found ways to use them to calculate the masses of their stars.

Based on what we have covered so far, what would we expect the above-mentioned plot to look like? This sounds like a weird question, but sometimes it's useful beforehand to figure out what you expect and then see how what really exists is like and unlike your expectations. In this way, you see your expectations in the light of day rather than letting them push you around unnoticed.

Suppose we took the height and weight of the inhabitants of a town and made a scatter plot of weight versus height. We would expect most of the points to be at small weight and small height, medium weight and medium height, or large weight and large height. We expect this partly because children have both smaller weight and height than adults; but also because adults come in different sizes and that for a given body shape, an adult of larger size is both taller and heavier. Our expectation is then that we should be able to draw a curve going from small height and weight, through medium height and weight, to large height and weight, and that most of the points will be fairly close to this curve.

The same sort of thing happens for stars, as shown in the H-R diagram given in figure 7. There is a curve, called the main sequence, going from low temperature and luminosity, through medium temperature and luminosity, and then to high temperature and luminosity. Most of the stars on the H-R diagram lie near the main sequence.

Why is this? Stars spend most of their lives burning hydrogen, so most of the stars that we see are stars in the hydrogen-burning part of their lives. However, hydrogen-burning stars differ from star to star in the amount of mass that they have since they did not all start out with the same amount of gas. A more massive star has more gravity to overcome, so it must generate more pressure. It does this by burning fuel faster. That is, it has a higher luminosity. The light comes to us from the star's surface, so a larger luminosity requires a larger surface temperature or a larger surface area or both. In the case of hydrogen-burning

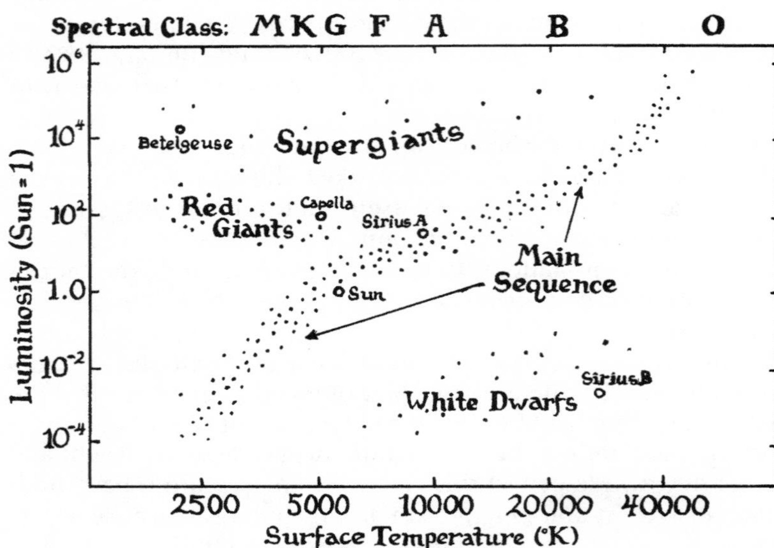

Figure 7

stars, it is both. This means that for hydrogen-burning stars there is a definite relationship between the two axes of our graph. In mathematical terms, such a relationship is called dependence. Luminosity "depends" on temperature because both depend on mass. This gives us the reason for the main sequence, since it shows the distribution of hydrogen-burning stars—or rather, of stars that are hydrogen burning when we see them, since at later times in their lives those stars will be very different in their temperature and luminosity than they are now.

"Main sequence" seems like a very odd name for the curve of hydrogen-burning stars. Just as odd is the fact that, in contrast to our figure 7, the H-R diagrams in astronomy textbooks have the temperature scale "backward" with high temperatures on the left and low temperatures on the right. What accounts for these notational oddities? They come about from an error in assumption. Russell thought that stars evolve from large hot objects to small cool objects, moving along the curve that he called the main sequence. If this theory were true, then the left side of the astronomy textbook H-R diagram would represent early times, the right side would represent late times, and the name "main sequence" would make sense as the sequence of states that a star goes through in its lifetime. We now know that Russell's theory is wrong. Instead, a given star during its hydrogen-burning time occupies approximately a single point on the main sequence, and the whole main sequence curve is made up of stars of different masses but in the same stages of life.

It would probably be less confusing if astronomers were to change the name "main sequence" to something else, and to use the orientation of the temperature axis used in this book. However, astronomers can be hidebound traditionalists just like anyone else, and as with OBAFGKM, names have a tendency to stick. Nonetheless, the name "main sequence" and the orientation of the temperature axis provide some interesting information about science: one of the main strengths of science is that only those theories that pass the challenge of experiments are kept. But this also means that scientists can be somewhat forgetful of the history of their discipline, not keeping in mind all the twists and turns and failed paths that eventually culminated in the successful theories that were kept. A term like "main sequence" is essentially a fossil from an early era and reminds us of a theory that was considered promising then but has not survived to the present day.

Most stars lie near the main sequence, but some do not. What about the stars far from the curve? Let's go back to the weight versus height curve for people. What do we call people who are well above the curve? We generally call them overweight, but why do we do that? It is true that they have a larger weight than the average person of their height. But it is just as true that they can be described as having smaller height than the average person of their weight; as the old joke puts it, "I'm not overweight. I'm under-tall." Most likely the reason for the usual terminology is that for adults height does not change appreciably, whereas weight can be changed by diet and exercise. In other words, we use the word "overweight" because it suggests to people that they might want to push themselves back into the more socially accepted regions of the graph.

A star well above the main sequence is both more luminous than the average star of its temperature and colder than the average star of its luminosity; do we call it "bright" or "cold"? Adding to the confusion is the fact that for a given surface temperature, the larger the surface area, the larger the luminosity. A star that is overly luminous for its temperature can simply be described as being large compared to a main sequence star of the same temperature.

The standard astronomy terminology uses all these properties and refers to stars well above the main sequence curve as red giants. Here "giant" is appropriate because the stars have a large size compared to main sequence stars of the same temperature, and "red" is appropriate because the star is colder (it's still very hot; cold is relative, after all) and therefore has light more toward the red end of the visible spectrum than the main sequence stars of the same luminosity. Similarly, stars well below the main sequence curve are called white dwarfs. Here "dwarf" is appropriate because such a star is much smaller than main sequence stars of the same temperature, while "white" is appropriate because such a star is much hotter and therefore gives off light that is much whiter than the very red main sequence stars of comparable luminosity. Red giants and white dwarfs have no means of dieting themselves back to the main sequence, but we doubt they're concerned about that; real stars don't care about their appearance.

From this matter-of-fact discussion of scatter plots and dieting, one might not guess how exotic and controversial white dwarf stars were when they were first discovered. That small size on the chart told scientists that white dwarfs were very dense. It may not sound very dramatic to say that a white dwarf is a star with the mass of the Sun and the size of Earth. However, this also means that a teaspoon of white dwarf matter has about as much mass as three elephants. There is no material on Earth even close to this density. It was perfectly just and cautious for scientists to be skeptical of the discovery of such a hypothetical material that could not be found anywhere nearby.

The first white dwarf discovered was Sirius B. The star Sirius is the brightest-appearing star in the sky. From observations of this star, German astronomer Friedrich Bessel concluded in 1844 that Sirius is a binary system: a bright star (Sirius A) that he observed, and a dim star (Sirius B) that he did not observe but whose presence he deduced from its effect on the motion of Sirius A.

Bessel is most famous for the mathematical functions that bear his name (these functions arise in the solution of many physics problems and must be learned by all college students majoring in physics, causing many such students to curse the name of

Bessel). He was the first to measure the distance to a star (other than the Sun) and was a pioneer in the development of very precise astronomical methods, including the analysis and reduction of error. He also found motions of the stars Sirius and Procyon and concluded that these motions must be due to the presence of unseen companions. In addition, he played a role in the discovery of Neptune by finding small irregularities in the orbit of Uranus and theorizing that these were due to an unknown planet beyond Uranus's orbit.

In 1862 Sirius B was found directly by Alvan Clark, an American telescope maker and astronomer. Sirius B's luminosity was found to be only about 1/1,000th that of Sirius A. This by itself is not surprising. A cold star is dimmer than a hot star; so astronomers simply expected that Sirius B had a low temperature. However, when Adams measured the spectrum of Sirius B, he found that the star was hot. Thus Sirius B was found to be a hot but very dim star. This meant that Sirius B had a very small size, calculated to be about the size of Earth. This, coupled with measurements showing that the mass of Sirius B was about 1 solar mass, implied the enormous density of white dwarf matter, which resulted in shocked disbelief from the astronomical community. As Eddington put it:

> The message of the companion of Sirius when it was decoded ran: "I am composed of material 3,000 times denser than anything you have ever come across; a ton of my material would be a little nugget that you could put in a matchbox." What reply can one make to such a message? The reply which most of us made in 1914 was—"Shut up. Don't talk nonsense."

However, eventually enough observational evidence accumulated (in the detected universe), and the theory of white dwarfs as parts of stellar evolution was worked out (in the theoretical universe). When these two universal views were firm enough, white dwarfs were accepted into the celestial family. In hindsight, the enormous density of white dwarfs doesn't seem so strange: the usual density of matter comes from the usual size of atoms, which are composed of nuclei and electrons. But because both nuclei and electrons are so much smaller than atoms, one can think of an atom as being mostly empty space. White dwarf matter is completely ionized; that is, all the electrons have been removed from the atoms forming a plasma: a gas of nuclei and electrons. Under the influence of the enormous gravitational field of a star, this plasma is compressed to the enormous density found in white dwarfs.

✷

White dwarfs are harder to observe than ordinary stars due to their small surface area. A white dwarf has about the surface area of Earth. A neutron star is even more extreme; it has a surface area comparable to that of a city. Neutron star matter, under the name "neutronium," is a staple of science fiction because of its huge density. A teaspoon of white dwarf matter has as much mass as three elephants, but a teaspoon of neutron star matter is about ten times as massive as all the people on Earth put together. However—sorry to disappoint science fiction fans—neutronium is not a material that one could work with like steel or plastic. It has its enormous density only because it is kept that way by the enormous gravitational field of the neutron star. A teaspoon of neutron star matter in empty space would quickly be blown apart by its pressure, and its neutrons would undergo beta decay, turning into protons, electrons, and antineutrinos. Soon after it was spooned out, there would be no sign that the neutronium had ever been there.

One might expect that it is extremely difficult to observe neutron stars. Indeed, J. Robert Oppenheimer and George Volkoff worked out the theory of neutron stars in 1939 but did not expect to be able to observe one. Though Oppenheimer is best remembered as the head of the atomic bomb project, he also made fundamental contributions to astrophysics. In addition to his work on neutron stars, Oppenheimer and his student Hartland Snyder did the first calculation of the process of black hole formation by gravitational collapse.

Despite the fact that neutron stars are not observed by the methods that work for ordinary stars and white dwarfs, they can be detected by other means. There are, as it turns out, several different ways to detect neutron stars using methods that relate to the consequences of what they are and how they behave. The most dramatic of these consequences concerns a kind of stellar object known as a pulsar.

Pulsars were discovered in 1967 by the astronomers Jocelyn Bell and Antony Hewish, who were using a radio telescope in Cambridge, England, to find the effect of solar wind on the observation of distant radio sources. They found a particular radio source that consisted of a series of regular pulses, a flash of radio signal that occurred exactly once every 1.337 seconds. It occurred to Bell and Hewish that these signals might have an artificial origin or, as Bell put it, "I was now two and a half years through a three-year studentship and here was some silly lot of Little Green Men using *my* telescope and *my* frequency to signal to planet Earth." However, they then found another source of regular radio pulses in the same frequency coming from a different part of the sky, which quickly put an end to the "Little Green Men" hypothesis. As Bell explained, "It was highly unlikely that two lots of Little Green Men could choose the same unusual frequency and unlikely technique to signal to the same inconspicuous planet Earth!"

It was eventually hypothesized that these pulsars were neutron stars. But why would a neutron star emit regular radio pulses? There are two contributing factors: the neutron star's very large magnetic field and its very rapid spin. Consider by way of metaphor how a lighthouse looks to a ship at sea during the night. The lighthouse light shines steadily, but its direction changes, rotating in a circle. A sailor on the ship sees the light when it is pointing at him, but not at other times. The sailor sees the lighthouse light as a set of regular pulses, one for each time the light goes around. So *if* Bell's pulsar is spinning around once every 1.337 seconds and *if* it is emitting radio waves only in certain directions, then that would account for the observations. But why would neutron stars spin so rapidly? And why would they emit radio waves only in certain directions and not others?

To understand this, we need to turn to an aspect of the Sun that we ignored: the fact that it's spinning. From the Sun's rotation, we will be able to determine characteristics of stellar rotation. The Sun rotates about once every twenty-seven days (as seen from the motion of sunspots), and so we might expect that that is a typical rotation rate for stars. However, neutron stars form from the collapse of a star. When spinning objects get smaller, they tend to spin faster. This principle, called the conservation of angular momentum, is familiar to figure skaters who pull in their arms and legs in order to make themselves spin faster.

Conservation of angular momentum is one of several such conservation laws in physics. Conservation laws concern quantities that cannot be created or destroyed, only transformed and transferred. The most famous/important of these is the conservation of energy, which states that energy is not created or destroyed, but only moved around from place to place or changed from one form of energy to another. Conservation of energy allows physicists to solve a great many problems in that they can always ask the question "Where did the energy go, or where did it come from?" Another important law is conservation of momentum. Momentum is mass times velocity. A large slow object can have the same momentum as a small fast object. Conservation of momentum is a little subtle since velocity is directional, which explains why a jet or rocket can move in one direction by shooting hot gas in the opposite direction. (The total of the mass of gas times the velocity in the direction of the gas motion and the mass of rocket times the velocity of the rocket in the opposite direction is zero; that is, a whole lot of work is done and the gas and the rocket each gain momentum, but in the whole system the momentum is zero.)

Angular momentum is like momentum but not quite; it has to do with mass, size, and angular velocity (which way and how fast the object is spinning). In any case, the figure-skater metaphor does a good job of showing how applying the conservation of angular momentum to the formation of a neutron star in the collapse of the core of a star yields a great increase in rate of rotation, and if calculated out, produces rates that are appropriate for pulsars.

Just as collapse tends to increase rates of rotation, it also increases magnetic field. This is perhaps best seen as a consequence of the approximate constancy of magnetic flux (it tends to stay the same, although it isn't conserved). Magnetic flux is the value of the magnetic field multiplied by the surface area. Since flux stays roughly the same, as the surface area of the neutron star shrinks, the magnetic field will grow. In the case of a star that has a strong magnetic field and a large radius, shrinking the radius makes that field even stronger. Thus a neutron star, an enormously concentrated core of a star, ends up with a very rapid rotation rate and a very large magnetic field.

How does this give rise to a radio signal that emerges from the neutron star only in a certain direction? It's actually the same effect that makes the aurora borealis: the motion of charged particles in a magnetic field. In this case, not only the charged particles but also the magnetic field are doing the moving because as on Earth the "true" and "magnetic" north poles of a neutron star are not the same things.

We learn as children that a compass points north. But long ago sailors learned to make the distinction between magnetic north (the direction the compass points) and true north (the direction pointing to the pole that Earth rotates around). On Earth the magnetic north pole and the true north pole are fairly close, so the rule that "the compass points north" is pretty good. However, these poles need not be close together. Since true north and magnetic north do not exactly coincide, once a day the magnetic north pole of Earth rotates in a circle around the true north pole. This creates a very weak signal from Earth, which has a weak magnetic field. But in the case of a strong field such as a pulsar, huge electric and magnetic fields are created as the magnetic pole rotates around the axis of rotation of the neutron star.

We went a little fast there. We need to explain why radio waves are emitted in the directions of the magnetic poles. Recall that the aurora borealis happens in the north because charged particles travel in the direction of the magnetic field. Also note that moving magnetic fields make electric fields. (This is the principle behind the generation of the electric power that we use in our homes.) The very large magnetic field and rapid rotation of the neutron star give rise to an enormous electric field that accelerates protons and electrons along the magnetic field of the neutron star. These accelerated charged particles produce the radio waves that form the observed signal of the pulsar.

Since the deduced magnetic structure of a neutron star would produce the effects noted by pulsars, scientists came to the plausible conclusion that this was the source of the pulse in pulsars, and that pulsars were not an unknown kind of object, but a confirmation of the hypothesized neutron stars. In other words, pulsars acted like neutron stars should act, so it was concluded that they were neutron stars. No little green men need apply.

Though pulsars are the most dramatic manifestations of neutron stars, they are neither the simplest detectable aspect of these old stars nor the most powerful. The simplest effect comes from the fact that a neutron star is a spinning magnet; and spinning magnets give off electromagnetic radiation in all directions. How do we notice this radiation? Let's look at one specific example: the Crab Nebula. In 1054 a supernova was observed by the Chinese and probably also by the Anasazi Native Americans. "Was observed" is actually too mild a phrase for a star suddenly lighting up visible to the naked eye and then going dim. Why don't we say that the star exploded in 1054? Because it didn't. The supernova is about 6,000 light-years from us, which means that it takes light about 6,000 years to get from it to us. Thus the Crab supernova was observed in 1054; but the star blew up about 6,000 years before that. In astronomy as noted above, the ability to look out into space is also an ability to look back in time. And, yes, we can look all the way back in time, as we will see in the next step.

A pulsar can now be detected at the same position in space as the Crab supernova, which helps confirm the idea that supernovae quickly (on the stellar scale) become neutron stars. This is one of the few cases where there were observations made long enough ago to serve as data in stellar evolution. In most cases, stars take too long to change for any human-scale observation to be useful in confirming changes in a single star. Supernovae are the exception since their explosions actually happen on a human time scale.

As we said, in a supernova, the core collapses and the envelope is ejected. In this case, the core has become a neutron star, but what has happened to the envelope? That remains as an ever-expanding cloud of gas that we see today as the Crab Nebula. This nebula is quite bright, shining with about 80,000 solar luminosities (that is, the Crab Nebula puts out about 80,000 times as much power as the Sun). If the supernova was observed about 1,000 years ago, why is the Crab Nebula still shining? What gives it its luminosity, since it is too diffuse to be fusing anything?

The nebula, strange as it sounds, is powered by the neutron star. The electromagnetic radiation given off by the spinning neutron star accelerates electrons. When electrons move in a magnetic field, they give off electromagnetic radiation, called synchrotron radiation. It is this synchrotron radiation that we observe when we see the Crab Nebula shining.

But where does the neutron star's energy come from? After all, it is no longer a powered star. There is no fusion going on in that mass of neutronium. It is not enough to say that spinning magnets make electromagnetic radiation. Energy is conserved, not created or destroyed, but rather moved around and transferred from one type to another. So why does a neutron star, which is not burning any fuel, have energy to emit?

There is energy of rotation in the spin of the neutron star, and this energy is powering the radiation. Ultimately, the energy that the neutron star emits leads to its gradually slowing down, and indeed very accurate measurements of the period of rotation of the Crab pulsar reveal that it is slowing down at just the rate that one would expect from the rate of energy emitted by the Crab Nebula. Thus by looking back to thousand-year-old records and looking at a beautiful bright image in the sky, we gain more confirmation of the existence and character of neutron stars.

The last method that we will consider for detecting neutron stars has to do with binary star systems. For our purposes, this is the most important detection method because it can also be used to detect black holes. When the stars of a binary system are far from each other, each holds on to its own material using its own gravitational force. However, when the stars are close to each other, it may happen that one star's gravity is strong enough to pull material away from the outer layer of the other star.

Consider such a close binary star system that has a neutron star and an ordinary star, which we will call the neutron star's companion. What happens to the material that the neutron star pulls away from its companion? It is torn from the companion by the enormous gravity of the neutron star and then guided by that gravity into orbit around the neutron star, forming a pancake-shaped object called an accretion disk. The accretion disk is continually depleted as gas from its inner edge falls onto the neutron star, while at the same time it is continuously replenished with gas from the companion. This process has some similarity to the Kelvin-Helmholtz mechanism that first heated the Sun. The gas is pulled in by gravity and thus compressed. The compression leads to heating, which in turn leads to the gas giving off light.

Let us consider, as we did for the Sun, how much energy per kilogram of fuel is released in this process. In the analysis of the Sun, we compared all fuels to the natural gas that some of us burn in our furnaces. Here we will compare the burning process to the theoretical maximum given by Einstein's formula $E = mc^2$. This familiar but misunderstood formula gives, among other things, the maximum energy that can be extracted from a hunk of matter. Since c is a large number (3×10^8 meters/second), c^2 is very large (9×10^{16} meters^2/second^2). Thus the ratio of maximum energy to matter is quite high.

When the electric power company charges us for energy, they measure it in kilowatt-hours, where a kilowatt-hour is the amount of energy that it takes to light a 100-watt lightbulb for 10 hours. Burning natural gas gives energy of about 15 kilowatt-hours for each kilogram of fuel. But Einstein's formula says that every piece of matter has locked within it an enormous amount of energy, about 25,000,000,000 kilowatt-hours for every kilogram of fuel. Let's define the efficiency of a fuel to be the amount of energy per kilogram that it delivers divided by this theoretical maximum. Then natural gas has an efficiency of only about .0000000006 (or 6×10^{-10} in scientific notation). Similar numbers apply to gasoline, so it is no wonder that we have to fill up our cars so often since we use such inefficient fuel. Hydrogen fusion does much better, with an efficiency of about .007. Fusion, the James Bond of energy sources.

However, the gravitational potential energy that the gas of the accretion disk releases does even better, with an efficiency of about .3 (or 30 percent of the maximum). In other words, the neutron star takes material from the companion and produces from it about 40 times as much energy as would be released if that same material had undergone nuclear fusion. This gives us the curious fact that a "dead" star can produce much more energy than a "living" one.

What does this release of energy look like? The gas is heated to a very high temperature as it falls toward the neutron star. Heated gas gives off electromagnetic radiation, and the hotter the gas, the smaller the wavelength of the radiation given off. Ultraviolet rays have a shorter wavelength than visible light, and X-rays shorter still. Neutron star accretion disks are so hot that they give off X-rays. Unfortunately, Earth's atmosphere tends to absorb X-rays. In fact, the atmosphere absorbs most wavelengths of light, with visible light and radio waves being the exceptions. This is actually only unfortunate for astronomers. For the rest of us, this absorption is good, as it protects us from some nasty radiation, but scientists tend to resent things that interfere with detection, even if it has other beneficial effects. To get around this "problem" without exposing us all to imminent death, X-ray telescopes have been put on satellites. These have been used to detect the accretion disks around neutron stars. The results of studying the X-ray sky revealed a number of objects whose X-ray emissions matched those expected in neutron star–normal star binary systems.

We have detailed several different ways in which neutron stars have been detected, but we have so far described these methods using the language of the theoretical universe. We have said that there are neutron stars and that they can be observed in several ways. However, we could instead use the language of the detected universe. We could say that several astronomical phenomena have been observed called pulsars, bright nebulae, and X-ray binaries. For each of these phenomena, there is a model and each model makes use of a theoretical entity called a neutron star. The advantage of this detected universe language is one of sensible caution. Pulsars, in the sense of regular pulses of radio waves from some astronomical source, have been observed. To the extent that the observations are reliable, we know that pulsars exist regardless of what we do or don't know about their nature. In contrast, calling a pulsar a neutron star commits us to a certain model of how a pulsar works and makes us wrong if that model is wrong.

The disadvantage of the detected universe language is that it both reflects and promotes a certain lack of understanding. This can best be illustrated by the fable of the blind men and the elephant. In this story, several blind men encounter an elephant. One man feels the trunk and says, "The elephant is like a snake." Another feels a leg and says, "The elephant is like a tree." Yet another feels the tusk and says, "The elephant is like a spear." This fable is humorous because we know the truth: the elephant is like all of these things and none of these things. Each blind man, however accurate his observations, has only a part of the truth. Only we with our sight (or the blind men if they could stop quarreling long enough to combine their observations) have a good picture of the whole elephant.

Typically in astronomy, the type of language used for certain phenomena changes over time. The observations come first, so that the language of the detected universe is forced on us, giving detection-based names such as "pulsar" to detected phenomena. Theories are proposed, and the language of the theoretical universe appears (neutron star, in this case), though sensible caution keeps it from becoming dominant. Finally, the theory becomes widely accepted. Then its language is prevalent with everyone, except possibly the observers who are sometimes fond of their own terms. Today neutron stars are in the last of these stages; the theoretical universe terms and explanations are broadly accepted and taking over from the old terminology (there may come a time when "pulsar" as a term is completely archaic). Black holes as a subject are somewhere between the middle and the last stage, mostly accepted but still having some artifacts and older views in use. This is partly due to the nature of the evidence, but is probably also due to a certain excess of caution on the part of astronomers. Of course, one person's excess of caution is another's common sense, and one person's clear-headed foresight is another's reckless zeal for their own views.

Rather than argue whether changing views of them is sensible or overzealous, let's just see if we can go out into the wilds of the universe and bag some black holes.

Don't worry: they're not an endangered species.

CHAPTER 7

Black Hole Hunting

Now that we have tracked down some furtive quarry by seeing how white dwarfs and neutron stars can be detected, it's time to try and find some black holes, which is what we were trying to bag in the first place. (Not that you want a black hole in a bag. It's bad for the bag, and for you.) We begin by considering the similarities and differences between black holes and neutron stars and seeing if the means used to find one hard-to-find object can help us find another. In the detected universe, it's often a good idea to see if you can adapt the tools made for one purpose to another, rather than having to build them from scratch. But in order to adapt the tools, it's necessary to make sure that the objects being worked upon are similar enough in the appropriate ways.

Both neutron stars and black holes are compact objects with large gravitational fields. However, unlike a neutron star, a black hole does not theoretically have a magnetic field—which eliminates two of the three neutron star–finding methods right away. Both pulsars and bright nebulae depend on the neutron star magnetic field for their detection. But, fortunately, the neutron star explanation of X-ray binary systems does not depend on the magnetic field since the X-rays that emerge from matter falling into a neutron star require nothing but a binary system and gravity.

We should expect that there are X-ray binaries where the object with the large gravitational field is a black hole. The tricky question here is how do we tell the difference between a neutron star X-ray binary and a black hole X-ray binary? At first one might expect that since a black hole has a more extreme gravitational field, it would have a larger fuel efficiency than a neutron star. Indeed, theoretical calculation shows that one could attain a fuel efficiency of 100 percent (all the matter converted into energy) by slowly lowering an object to the black hole event horizon. However, mathematical models of accretion disks around black holes show that the disk does not extend all the way to the event horizon. Instead, there is an inner edge to the disk from which the gas plunges quickly through a gap into the black hole. The fall is not slow enough for increased efficiency. Not all of the energy from the accretion disk's lost matter is radiated away in X-rays. Some of it falls into the black hole. Taking this effect into account, the fuel efficiency of a black hole accretion disk is not very different from that of a neutron star accretion disk. No help there.

However, we know that black holes, unlike neutron stars, do not have a maximum mass. A neutron star that becomes too massive becomes a black hole; a black hole just becomes a bigger black hole. Thus in an X-ray binary, if the mass of the compact object is greater than the maximum neutron star mass, we can conclude that the object is a black hole. A handful of black holes have been identified in this way. Unfortunately, this is somewhat circular reasoning; what we are saying is that if we spot something that should be a neutron star but is too massive, it has to be a black hole. This is using theory, not detection, to distinguish two objects and is a cause for caution. After all, there might be some unknown way in which a neutron star could avoid collapse despite its larger mass. We need another method.

Let's think again about the differences between these two collapsed stars. We know that black holes differ from neutron stars in that neutron stars have a surface and black holes have an event horizon. From the inner edge of the accretion disk, therefore, matter falls onto the surface of a neutron star in a neutron star X-ray binary. But that same matter falls through the event horizon in a black hole X-ray binary. The two types of binaries can be distinguished provided that we can reliably tell the difference between their behavior in accreting matter. In other words, if we can tell the difference between matter falling on something and matter falling into something, then we can tell neutron stars from black holes.

Unfortunately for reasons of modeling difficulty discussed below, we can't find a reliable method for doing this with small black holes. This is disappointing. Black holes of a few solar masses have been detected in X-ray binaries, but only in small numbers and even then with a bit of uncertainty as to whether what has been found is a black hole or a neutron star.

However, the situation changes when we consider more closely the fact that a neutron star has a maximum mass but a black hole does not. A neutron star must have less than about 2 solar masses. But nothing prevents a black hole from having thousands or millions or even billions of solar masses. Such a supermassive black hole is nothing like a neutron star, and its accretion behavior would stand out for millions of light-years. In other words, if we aim bigger, if we look at objects much larger than stars, we might find our black holes without any risk that they are neutron stars.

So we are led back to a variation of our first question in this chapter, a variation on the question of Michell and Laplace: Do supermassive black holes exist? Both those researchers pointed out that 100 million solar masses at the density of water would form a black hole. At first sight the density doesn't seem extreme, but the amount of mass does. Where are you going to find 100 million solar masses in a small-enough volume to form a black hole?

The answer can be found by moving up from the scale of stars and considering the properties of galaxies. Stars, despite the appearance of the night sky, are not spread uniformly through space, but instead are "clumped" in huge collections of stars called galaxies. How and why they are so clumped is (at least partially) known, but we will postpone a discussion of that until the next chapter.

Our own galaxy, the Milky Way galaxy, is a collection of billions of stars in a disk about 40,000 parsecs in diameter. "Milky Way" is what a huge disk of stars looks like to someone living in that disk. Since a disk is thin in one direction and thick in others, someone living in a disk of stars would see few stars when looking in directions out of the plane of the disk, but many stars when looking in the plane of the disk. A "band" of directions in the sky would have a diffuse brightness created by the light of the many stars. This band was easily visible to ancient astronomers, who named it the Milky Way. The word "galaxy," incidentally, comes from the Greek word for milk.

Though 100 million solar masses is not a large mass on the scale of galaxies, the density of water is a very high density when you consider interstellar distances. To see this, note that in our "neighborhood" of the galaxy, stars are a few parsecs apart, so the density in stars is about 1 solar mass per cubic parsec. Since the radius of the Sun is much smaller than a parsec, this means that the density in our neighborhood of the galaxy is much smaller than the density of water.

Nonetheless, the center of the galaxy is denser with stars than the outer regions, and our Sun is far from the center of the galaxy (we live in the galactic sticks, far from the hip downtown action at the center of the Milky Way). It is not implausible that the centers of galaxies become dense enough to undergo gravitational collapse and form black holes. However, our understanding of the process of galaxy formation and subsequent development is not good enough to make a definite prediction about the presence or absence of black holes in the centers of galaxies. Instead, the question is one of detection. Are there supermassive black holes in the centers of galaxies?

We begin with a peculiar truth that follows from the discussion of accretion disks: while black holes are black, if they are near something they can eat, they give off a lot of energy. In other words, around the darkness is a huge amount of light (particularly X-ray light). We are asking whether there are supermassive black holes at the center of galaxies, places full of juicy stars to eat. In such a galaxy, we would expect to detect a very bright center.

A galaxy with a bright center was first found as far back as 1908 by the American astronomer Edward Fath using the telescope of the then newly constructed Mount Wilson Observatory in Southern California. However, the first systematic study of such galaxies was done by the American astronomer Carl Seyfert in 1942, also at Mount Wilson. These are referred to as active galactic nuclei or by the acronym AGN (no clever pronunciation, sometimes an acronym is just an acronym). Here the word "nuclei" has nothing to do with the nuclei of atoms but simply means centers. "Nucleus," whenever it is used in science, generally means center or central object, so we have the nucleus of an atom, the nucleus of a cell in biology, the nucleus of a galaxy, and so on. The spectrum of light from an AGN typically peaks in the ultraviolet, but they also emit X-rays as well as visible light and infrared. In addition, there are bright emission lines. These are bright spectral lines from a hot diffuse gas. About 90 percent of AGNs emit little in the way of radio waves and are called radio quiet, while the other 10 percent emit copiously in the radio frequencies and are called radio loud.

Perhaps the most dramatic of the AGNs are the so-called quasi-stellar objects, usually known by the acronym QSO. These were first detected in the 1960s by Thomas Matthews and Alan Sandage, who were searching for optical counterparts to radio sources. Basically, astronomers using radio telescopes had found certain objects that emit radio waves, and Matthews and Sandage using ordinary (optical) telescopes wanted to find out whether those objects also emitted visible light. What they found was at first sight surprising: a set of bright emission lines that did not seem to correspond to any known chemical element. This was interesting but did not, unlike the similar confusion about helium, lead to a discovery of a new element. In 1963 the Dutch astronomer Maarten Schmidt noticed that the mystery lines of a similar object were simply those of hydrogen, but subjected to a very large redshift corresponding to motion of about 15 percent of the speed of light.

As we will see in the next chapter, the expansion of the universe means that galaxies are moving apart from each other, and the farther the distance between the galaxies, the faster their relative motion. A galaxy moving away from us at 15 percent of the speed of light is more than a billion light-years away from us. The QSOs were at intergalactic distances, so the resolution of the mystery of the spectral lines led to another mystery: How could such an object even be visible? To be seen at that large distance, an object would have to be emitting a huge amount of energy. The Milky Way is a typical galaxy; but a typical QSO has about 500 times the luminosity of the Milky Way, while the most luminous of the QSOs have a luminosity of about 100,000 times that of the Milky Way. So what could possibly be generating so much energy as to dwarf the output of billions of stars?

The mystery deepened when astronomers considered the variability of the QSOs. Each particular QSO changes how bright it is over short periods of time (short for humans, ridiculously short for stars). But things that change brightness do not appear to do so all at once across their entire surface. For an object of a certain size, even if its brightness changes all at once from its point of view, we will see the brightness changing at different times on different parts of the object, simply because the light from different parts of the object reaches us at different times. For example, if an object 1 light-minute across dimmed to half its brightness at once, we would see the closest part dim 1 minute before the farthest part did. This actually happens even on the human scale, but the size of objects we are used to is so small compared to the speed of light as to make the changes seem instantaneous. We think we see a lightbulb blow out in an instant, but in reality the far end of the bulb changes (from our perspective) about 5 centimeters (lightbulb radius)/30,000,000,000 cm/sec. (speed of light in centimeters per second) = 1/6,000,000,000 sec. or one–six billionth of a second after the front changes (okay for our perceptions that's an instant, but the reality of the time lag is important).

If an object has a size of 1 light-year, then the most rapid variability (time scale on which the brightness of the entire object could change) is 1 year. However, typical QSOs have been observed to have a variability of a few hours, meaning that they had to be at most a few light-hours across. In other words, QSOs had to be smaller than the solar system while putting out more power than an entire galaxy.

Furthermore, the large QSO luminosity had to correspond to a large mass. Recall that a star is held in equilibrium by the balance between gravity and pressure. This also holds for any gravitationally bound system. This means that for a given mass, there is a luminosity (called the Eddington luminosity) such that the object can't have a luminosity greater than the Eddington luminosity or the energy of the light would generate enough pressure to blow the object apart. Turning this logic around, a gravitationally bound object (that is, an object that has not exploded under its own power) with the luminosity of a typical QSO must have a mass of at least about 300 million solar masses in order to hold itself together. To summarize, a typical QSO has a luminosity of about 500 typical galaxies, a size smaller than the solar system, and a mass of at least 300 million solar masses. What could it be? Might there be a hypothesized kind of object that would fit this description? Hmm, what could it be?

Astronomers often phrase this last question as "What is the 'central engine' that powers QSOs and AGNs?" To that they add other questions: Why are there no nearby QSOs? Or turned around: Why are the QSOs more luminous than any of the nearby galaxies, even those with AGNs? Why do only some galaxies have AGNs? What generates the radio waves in radio-loud AGNs and radio-loud QSOs? By the way, radio-loud QSOs are often called quasars. It turns out that all these questions can be answered by the theory that the center of each and every galaxy contains a supermassive black hole with an accretion disk. (Which means that we should be happy living away from the hip center of the galaxy. The crush downtown is nasty.)

Recall that the Schwarzschild radius of a black hole is proportional to its mass and that the Sun has a Schwarzschild radius of about 3 kilometers. This means that a black hole with a 300 million solar mass has a Schwarzschild radius of about 900 million kilometers or about 6 AU. This is smaller than the solar system and within the limits set by the variability of QSOs and AGNs. With a fuel efficiency of about 10 percent, the luminosities of QSOs could be maintained by accreting about a few solar masses of material per year. The less luminous AGNs could generate their power with a correspondingly smaller accretion rate. This last point leads to the answer to the question of why QSOs are more luminous than AGNs. QSOs emit more power because they accrete more matter.

At first sight, this answer only seems to shift the mystery to another mystery, since we are still left with the question of why the black holes in faraway galaxies are accreting matter at a rate greater than the black holes in nearby galaxies. But remember the farther away an object is, the longer it takes light to get from it to us. So the greater distance of the QSOs means that the events we are witnessing in them happened farther back in time than the AGN events. We are seeing the QSOs and AGNs not as they are now, but as they were when they emitted the light that we are seeing now. Since QSOs are farther away than AGNs, we are seeing them at an earlier stage in their lives. Thus the difference between QSOs and AGNs simply reflects that supermassive black holes eventually use up their fuel. At earlier times they are consuming fuel more rapidly and burning more brightly; while at later times they consume fuel less rapidly and are more dim. This also explains why most galaxies today (that is, the galaxies that we can see close to us) do not have AGN. Their central black holes have used up all their fuel and have become relatively quiet. Notice that just as looking at different stars gives us an understanding of stellar evolution, looking at QSOs and AGNs gives us some understanding of galactic evolution. We will touch on this again in the next chapter.

What about all the different types of radiation given off by the QSOs and AGNs, the visible light, ultraviolet, X-rays, and radio waves? These, too, can be explained by a supermassive black hole with an accretion disk. The accreted matter gets heated up and emits energy because it's hot, just as happens in the case of neutron stars and stellar mass black holes. One might expect that the much larger masses of supermassive black holes would lead to a larger temperature of the accretion disk. But a supermassive black hole also has a much larger size; thus while its accretion disk radiates more energy, it does so from a larger area. The temperature is related to the power per unit area (the amount of power passing through each little bitty piece of the surface); and when all this is taken into account, it turns out that the temperature of the accretion disk is smaller for a supermassive black hole than for a stellar mass black hole (because the giant surface area more than makes up for the giant mass), and the radiation from the hot gas of the accretion disk is most prominent in the ultraviolet for the supermassive black holes as opposed to the X-ray prominence of neutron stars and stellar mass black holes.

The energy we see, however, does not just come from interaction with the black hole itself. One must also consider the behavior of the matter in the disk. Though a black hole has no magnetic field of its own, the plasma of the disk, through its rotation, can generate its own magnetic field. Through the rapid rotation, this magnetic field (as in the case of the neutron star) produces an electric field that in turn produces an electric current in the plasma. This combination of electric and magnetic fields and electric currents can extract rotational energy from the hole in the form of electromagnetic radiation and accelerated electrons, as first theorized by astrophysicists Roger Blandford and Roman Znajek in 1977. In other words, the accretion disk is not just feeding the black hole; it is also stealing its rotational energy, slowing it down, and in the process spitting out radiation.

In all of this activity around a black hole, there is a complicated and ongoing transfer of energy between gravity, plasma, and electromagnetic field. Moving plasma makes magnetic fields. Changing magnetic fields make electric fields. Electric fields accelerate electrons in the plasma. Collisions between electrons and

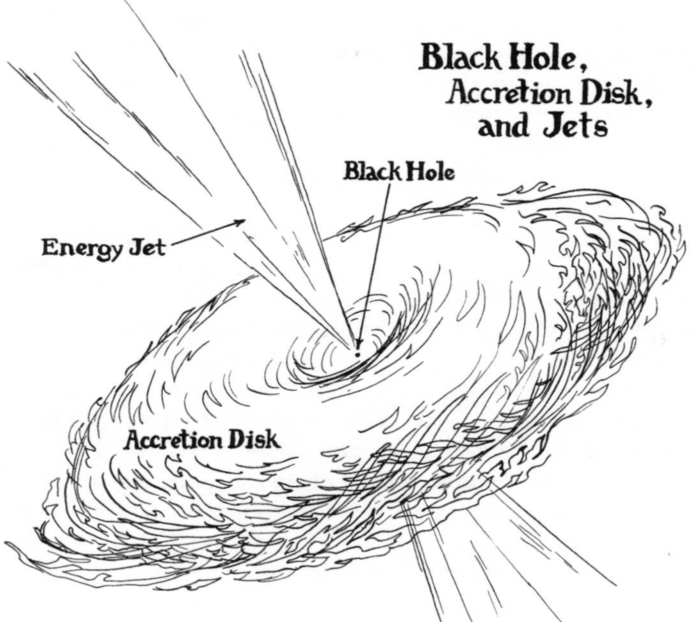

Figure 8

photons transfer energy from one to the other. Electrons moving in a magnetic field emit synchrotron radiation. A peculiar dance of energy transfer and energy forms is going on that produces a subtle and complex spectrum—but not a uniform spectrum. One might expect all this energy to come out equally in all directions, but it turns out that two effects sometimes make much of the energy come out in two jets perpendicular to the plane of the accretion disk.

The first effect has to do with the disk itself. All this radiation tends to swell up the inner part of the accretion disk so that it is shaped more like a doughnut than a pancake. (Here the position of the black hole is in the middle of the doughnut hole; see fig. 8.) The energy is then preferentially emitted in directions that don't collide with the doughnut—that is, in two jets out of the plane of the doughnut—since energy that collides with the doughnut is absorbed back into the doughnut. These jets are especially prominent in radio-loud AGNs and quasars.

The second effect has to do with the magnetic field of the disk. Plasma, since it is made of charged particles, tends to move in the direction of the magnetic field. The shape of the disk's magnetic field tends to focus the plasma in two jets at the poles of the field. Some of the radiation emitted is then absorbed by clouds of gas near the central accretion disk and by the outer parts of the disk itself. This tends to heat these objects and to bring their atoms to excited states. In particular, it is these excited atoms that then produce the emission lines that are characteristics of QSOs and AGNs. However, the electrons in the jets continue to move outward and continue to produce synchrotron radiation through their interaction with the magnetic field. This is what gives rise to the radio waves of radio-loud galaxies and quasars. Note that what is finally observed depends on the mass of the black hole, the rate at which it is accreting matter, and *the direction from which we observe it*. If we happen to be facing one of the jets, we will see an object that is seemingly more energetic than the same object would be if we were looking at it through the edge of the doughnut.

On Earth if we want to get a different perspective on something, we simply move around it. In this way we can see, say, a mountain from all sides and discover all its features. But we cannot in stellar and galactic terms shift our position. We are forced to see the pieces of the universe from one side only. In this way, observation of the wider universe is frozen, as a photograph is frozen. You cannot take a photograph and flip it around to see what's happening in the back. Just as you cannot see the back of a person standing face forward in a photo, so you cannot flip around a quasar to see a radio-loud AGN.

We have been a little quick in our analysis. Before we can proceed with care, we need a more clear theoretical model for what we are talking about. In the case of the Sun, we had a detailed solar model, taking into account all relevant physical effects and making quantitative predictions for the properties of the Sun. To create a theory that would fit with the detections of QSOs and AGNs, we would like to have a similar model for a supermassive black hole with an accretion disk.

It turns out that one part of developing such a model is easy and one is hard. Remarkably, the easy part is the black hole itself. The model for the characteristics and behavior of a nonrotating black hole was created by Schwarzschild (which is the reason for the name "Schwarzschild radius" for the size of the black hole) in 1916, just one year after Einstein first proposed his general theory of relativity. The formula for the rotating black hole took somewhat longer; it was done by the New Zealand mathematician Roy Kerr in 1963. It is a sign of the peculiarly simple nature of black holes that all the properties of a rotating black hole depend on only two quantities: the black hole's mass and its spin. Given the mass and spin of a black hole, Kerr's formula will tell us everything there is to know about that black hole.

In this sense, black holes are some of the simplest objects in nature, much simpler than the Sun and other stars, which are powered by subatomic processes and work through convection and have layers and all kinds of complexities. In some respects, black holes are simpler than the most basic subatomic particles, which tend to have more than just two quantum numbers. The idea that a huge macroscopic object like a black hole needs only two numbers is amazing and makes certain aspects of this theory easy.

Unfortunately, there is a little snag. In order to model the "central engine" of QSOs and AGNs, we need to model not just the black hole, but the properties of plasmas moving at high speeds and interacting in various complicated ways with electric and magnetic fields whose properties also need to be modeled. Add to all this the complicated structure of the strong gravitational field of a rotating black hole and a computational nightmare ensues. In other words, we need to model the doughnut, not just the hole. The fundamental forces of the universe are creating a truly messy dance around the simplicity of the black hole. Figuring this out is like trying to understand a hurricane by looking at the calm eye. Though much progress has been made in modeling the supermassive black hole, we are nowhere near having a complete model in the sense in which the solar model is complete.

Let's retreat a little from this complexity and look at the simple part. Black holes, as noted, are simple. They require only two numbers, mass and spin. We can know all about a black hole itself (as opposed to a black hole and its accretion disk) if we can discern those two numbers.

So far we have talked about estimating the mass of supermassive black holes only through estimates of what mass is necessary to hold together an object with the luminosity of a QSO. However, we can measure the mass of a supermassive black hole more directly using essentially the same method used for the

Sun. A supermassive black hole in the center of a galaxy has stars and clouds of gas in orbit around it, just as the Sun has planets in orbit around it. The same formula that allows us to use the speeds and distances of the planets to find the mass of the Sun can be applied to supermassive black holes.

For the black hole in the center of our own galaxy, we can track the motions of stars in close orbits around it. In this way, its mass has been estimated to be around 2.5 million solar masses. For other galaxies, we measure the velocity directly using the Doppler effect, and then using that and the distance, we find the mass. Using this method, the masses of many black holes in the centers of galaxies have been found.

Though black hole mass measurements are well established, the measurement of black hole spin is still in its infancy. This is because, though the gravitational field of the black hole depends on both mass and spin, the dependence on spin is very small except in regions close to the black hole. In principle, the black hole spin can be measured by detecting its effect on the innermost regions of the accretion disk. However, to do this accurately, the accretion disk would need to be modeled well enough to know quantitatively how its detected properties depend on the spin of the black hole. Preliminary measurements of this sort have been done with interesting results. It appears that black holes with accretion disks are spinning almost as fast as they possibly can.

✸

Throughout this book, we have emphasized the question "How do they know that?" as a way of understanding science. That question presumes that the part of science being discussed is in fact known. But scientific research is a continual effort to push back the boundaries of the unknown, so scientists are always working on the border between what is known and what is not known. Most of the material we have discussed so far is well understood. But for the rest of this chapter, we will discuss some of the more exotic and recently assayed properties of black holes, and thereby begin our journey into the only partly known realms at the frontiers of physics.

Perhaps the best known investigator of black holes is the British physicist Stephen Hawking. It is well known that, due to the disease ALS, Hawking's physical condition is gradually deteriorating, so that now he uses a motorized wheelchair to get around and a computer-generated voice to speak (leading some of his British colleagues to complain about "Stephen's American accent"). What is less often said is that over time Hawking's style of doing physics has also gradually changed. In the 1960s and early 1970s, Hawking was among the most mathematical of physicists, using the sophisticated geometric and topological techniques pioneered by British mathematician Sir Roger Penrose to prove theorems about black holes, the expanding universe, and the process of gravitational collapse. In the 1980s Hawking turned to more speculative matters: in collaboration with American physicist James Hartle, Hawking tried to apply quantum mechanics to the whole universe in an attempt to explain the origin of the big bang. In between these two periods, Hawking achieved just the right mix of mathematics and speculation to produce his masterpiece: the theory of black hole radiation, now known as the Hawking effect.

So far we have considered radiation produced by a black hole when matter falls on it. We would think, since black holes never let anything out, that a black hole left alone would be black, emitting nothing. From this view, it is startling that a black hole left to itself with no matter around it will still give off radiation. Indeed, in 1974 when Hawking first announced this result, it was considered quite surprising.

To understand the Hawking effect, it is best to start with the possibility of 100 percent fuel efficiency of black holes. An object slowly lowered to a black hole's event horizon converts all its mass to energy in the form of work done in the process of the lowering. This suggests that if we could somehow continue the slow lowering process past the point where the object crosses the event horizon, we could get even more work out of it—in other words, that the object would have a negative energy. This is not possible for ordinary macroscopic objects, which are modeled using classical physics, where energy is always greater than or equal to zero. But, remarkably, something like this is possible for individual particles seen from the models of quantum mechanics.

We're about to dive into one of the elements of modern physics that is abstract and hard to connect to normal thinking. In quantum field theory, the quantum mechanical theory of the properties of particles like photons and electrons, one of the most important objects in the universe is "nothing"—or rather, not so much "nothing" as a vacuum. In this region of the theoretical universe, the vacuum is not considered empty, but instead filled with "virtual particles." These virtual particles are created in pairs, one particle with positive energy and one with negative energy, and then a short time later this pair of particles annihilates each other. Since the net energy is zero, this annihilation is simply a return to the vacuum. From this view, you can imagine empty space as roiling and boiling with particles that exist for brief moments of time then disappear.

Now consider this process taking place outside of, but near, the event horizon of a black hole. Under those circumstances, it sometimes happens that instead of the pair of particles annihilating each other, the negative energy particle falls into the black hole while the positive energy particle escapes from it. This means that rather than the particles returning to nothing, they create a pair of effects. For those of us observers outside the event horizon, the net effect is that a particle has come to us from the vicinity of the event horizon, and the black hole's mass has decreased (because it has absorbed a negative energy particle and so its mass has gone down by E/c^2 where E is the energy of the negative energy particle). Although nothing can come from the inside of a black hole, nonetheless the net effect of this process is that the black hole is giving off radiation and its mass is decreasing. It's as if the universe is performing a trick where nothing leaves a locked box, but there ends up being more stuff on the outside and less on the inside.

It sometimes happens that both the positive energy particle and the negative energy particle fall into the black hole, yielding no net effect. However, it never happens that the negative energy particle escapes, since for real (as opposed to virtual) particles, negative energies are possible only inside the black hole.

Another surprise of the Hawking effect is the nature of the radiation given off: it is exactly that of an object that has a temperature. Hot objects give off more radiation than cold objects, and the radiation produced in this case acts exactly like radiation from the heat of a body. In general, small black holes are hotter than big black holes unless the small hole is much closer to its maximum spin than the large one.

At first it seems that the Hawking effect would be easy to observe, since we would only need to detect the radiation. But that hope is quickly dashed by a careful examination of Hawking's formula for the black hole temperature. The coldest possible temperature is absolute zero, which is approximately −273 degrees Celsius (−459 degrees Fahrenheit). Physicists measure temperature in degrees kelvin, which is the same as Celsius degrees except that kelvin measures degrees above absolute zero rather than degrees above the freezing point of water. To find degrees kelvin, take degrees Celsius and add 273.3. So water freezes at 273.3 kelvin and boils at 373.3.

Hawking's formula gives for a black hole of 1 solar mass the chilly temperature of about a ten-millionth of a kelvin (10^{-7} kelvin)—that is, a ten-millionth of a degree above absolute zero. For supermassive black holes, the situation is much worse, since larger-mass black holes have smaller temperatures. To make matters even worse, as we will see in the next chapter, the universe is filled with radiation at a temperature of about 3 kelvin. Since black holes are colder than this, this means that black holes are masked by this temperature (the same way a dim light is masked by a much brighter light). Also since there is this ever-present background radiation to feed them, black holes are absorbing more radiation than they are emitting.

Since the temperature is inversely proportional to the mass, this means that only a really small black hole would have an appreciable temperature. But as we pointed out before, at present it seems that there are no good observational or theoretical reasons to expect that such small black holes exist. Annoyingly, the same technique that allowed us to find really big black holes forces us to find only really cold black holes. Thus the Hawking effect, beautiful as it is in combining the grandeur of black holes and the weirdness of the quantum vacuum, is firmly stuck in the theoretical world, with no component in the perceived or detected worlds.

At first sight, this doesn't seem so bad. After all, there are certain parts of the solar model (like the temperature at a point halfway between the center and the surface) that we don't have direct observational evidence of, but which we accept because they are part of a theory that has been confirmed in other ways. But Hawking's calculation relies on a method called quantum field theory in curved spacetime, a way of doing the calculations of quantum field theory in the curved spacetime of general relativity. (We will discuss the curvature of the universe in the next step.) While there are several experimental tests of quantum field theory and several experimental tests of general relativity, there are no experimental tests of the combination, quantum field theory in curved spacetime. This produces the curious situation that we can confirm two sides of a theory but not the combination; we must tread with care when trying to mate these two beasts.

Is there then any reason why we should believe Hawking's calculation? Yes, sort of. We have some confirmation because we can get to the Hawking effect from two other sources: black hole thermodynamics and the Unruh effect. The latter is named for the Canadian physicist William Unruh, who has done much work on general relativity and on issues that arise from trying to combine general relativity with quantum mechanics.

Thermodynamics is the study of heat and temperature. It contains a concept called entropy, which is a mathematical measure of the amount of disorder in a given system, as well as a rule (called the second law of thermodynamics), which states that entropy always increases. Since that rule is the "second law," it is natural for the reader to wonder whether there is a "first law of thermodynamics" and if so, what it is. The first law of thermodynamics is that heat is a form of energy and that energy is never created or destroyed but simply changed from one form to another. So the first law of thermodynamics is conservation of energy and, by the way, heat is a kind of energy.

That things tend to disorder will not surprise anyone who has seen the inside of a teenager's bedroom. But it is perhaps surprising that disorder can be quantified and the statement that disorder increases made into a physical law. How this is done can be explained using the teenager's bedroom (yes, we're being prejudiced, but we both have children so we're using standard parental bigotry, which is socially acceptable, at least to other parents). In a neat room, everything is in its proper place (books on the bookshelf, clean clothes in the dresser, dirty clothes in the hamper, and so on) while in a messy room, things are not in their proper places ("What is that cheese sandwich doing on the computer? It's for the mouse, Dad").

Since each item has one proper place and many improper places, if the teenager doesn't care where things go and puts them down at random, then they are more likely to go in improper places and the room is likely to be messy. In other words, entropy tends to increase because there are more ways to be messy than there are to be neat.

The concept of entropy and the second law of thermodynamics have consequences that reach far beyond yelling "Clean that up" to teenagers. One example of entropy occurs in the commonplace freezing of water. Water becomes ice at 0 degrees Celsius (32 degrees Fahrenheit). In ice, the molecules of water are more ordered and more tightly bound to each other. If ice is more ordered than liquid water, then why is its formation not itself a violation of the second law of thermodynamics? We have to be a little more clear on entropy.

The second law of thermodynamics says that the *total amount* of entropy in a system increases; this means that we can reduce entropy in one part of a system as long as there is someplace in the system to dump the excess entropy. (This, by the way, is also why air conditioners work; excess entropy is dumped outside of the air-conditioned environment.) Since the molecules of the water are more tightly bound in ice, the ice has less energy than the liquid water. That extra energy is released to the objects outside the ice, where it creates disorder by moving their molecules faster. Ice forms precisely at that temperature where the extra disorder created by its release of heat can compensate for the order created by its formation.

A second example of the use of the second law of thermodynamics has to do with the efficiency of fuel use in cars. When a car runs, the engine gets hot. That is, some of the energy from the fuel is used to move the car, but some of it is wasted as heat. It turns out that well over half of the energy is wasted in this way. When first learning this fact, we might become incensed at the automotive engineers in Detroit (and Stuttgart and Nagoya) and want to exhort them to design a better product. But it's not completely their fault. A car engine that turned all the energy from its fuel into mechanical work is forbidden by the second law of thermodynamics. It would decrease the entropy. In fact, given the temperature in the engine cylinders (which is limited by the material that car engines are made of), a fuel efficiency of about one-half is the best that can be done. (Not that that efficiency is available. Other effects act to make the real efficiency of car engines lower than this ideal efficiency, nor are current efficiencies up to even this lower-than-ideal ideal.)

What does thermodynamics, which seems to be a totally unrelated branch of physics, have to do with black holes? Before we get to that, it is important to remember that the universe fits together and that universal rules apply in all circumstances. This means that we are perfectly justified in drawing on the second law of thermodynamics and putting it to the test in the realm of black holes. One of the great advantages of scientific thinking is knowing how and when to transport an idea or a principle from one thing to another.

Thermodynamics relates to black holes because the behavior of thermodynamic systems in the presence of black holes creates a crisis for the second law of thermodynamics. We dealt with the fact that we can't observe things that happen inside a black hole event horizon by saying that we would consider only phenomena that take place outside of black holes. Let's consider then what happens when an object falls into a black hole. As far as the world outside of black holes is concerned, that object has disappeared. At first it seems that this might create problems for the first law of thermodynamics. But recall that mass is a form of energy and that a black hole's mass can be observed from its gravitational effects. Thus the energy of the object is not lost when the object falls into the black hole; it is simply converted into black hole mass. As far as the conservation of matter and energy goes, we're fine.

But when the object falls into the black hole, its entropy really is lost. Thus it seems that the total entropy (at least the total entropy outside black holes) can decrease. It then seems that as long as there are black holes, we have an exception to the second law of thermodynamics (as well as a place to dump the mess from a teenager's room).

This difficulty is resolved by the Hawking effect. Since a black hole has a temperature, it is a thermodynamic system, and thermodynamic systems have entropy. In fact, it follows from Hawking's calculation that the entropy of a black hole is proportional to the area of its event horizon. (In fact, even before Hawking's work, it was suggested by the Israeli physicist Jacob Bekenstein that the difficulty black holes pose for thermodynamics could be resolved if black holes have entropy and if that entropy is proportional to the area of the event horizon.) Thus an object falling into a black hole does not violate the second law of thermodynamics. The object's own entropy disappears, but in the process the black hole's mass increases and therefore its event horizon area increases. When both of these effects are taken into account, it turns out that the total entropy increases. In other words, we can take heart, because the Hawking effect resolves a conflict in thermodynamics.

This is not a circular justification, since even before Hawking's calculation the rules for how black hole mass and area change when objects fall into them had been worked out. It had been noticed at that time that these "laws of black hole mechanics" bore a resemblance to the laws of thermodynamics. However, at the time it was thought that this was just a coincidence. After Hawking's calculation, it became clear that there are no separate laws of black hole mechanics. Black holes are thermodynamic systems, and the "laws of black hole mechanics" are just the ordinary laws of thermodynamics applied to black holes.

This use of thermodynamics as evidence for the Hawking effect can be regarded as only an aesthetic reason for believing a theory. The theory brings together gravity and thermodynamics in an unexpected way to form a coherent whole that looks really good to physicists. In the process, the theory resolves a paradox having to do with the compatibility of black holes and thermodynamics. In this sense, the theory is "too beautiful not to be true." It is sometimes claimed these days (especially by string theorists) that a particular theory is "too beautiful not to be true." However, it is then rarely spelled out wherein the beauty of the theory lies and why that beauty should be regarded as a compelling reason for acceptance of the theory. For the Hawking effect and black hole thermodynamics, these things have been made clear.

This aesthetic argument should create discomfort, since up to this point we have not resorted to standards of beauty for accepting scientific ideas; but there is more to this argument than simple appreciation. We can rephrase the argument as follows: Either the Hawking effect happens, or several basic strands of science that have been confirmed under other circumstances fall apart. This does not mean that we have to accept the Hawking effect, only that unless we find some other reason to mistrust thermodynamics, we should err on Hawking's side.

In science there are rarely individual exceptions to laws. When one finds an individual situation that seems to be an exception, one looks to see what broader circumstance would create that exception. In this case, we presume the validity of the second law of thermodynamics, since it has stood up to many tests. We know that it might not hold in these circumstances or might hold for some completely different reason than Hawking's, but until we have reason to think it is actually being violated, we accept it; and adjunct to it, we accept the Hawking effect, since it makes the whole fit together harmoniously. It may not be right, but its ability to hold together and explain everything gives us cause to accept it until and unless evidence points elsewhere.

Fortunately, we have another strand to draw upon to strengthen Hawking's hand: the Unruh effect. Unruh asked a general question seemingly unrelated to this subject. He asked what radiation, if any, is absorbed by an accelerating object passing through a vacuum. In other words, since the vacuum is really boiling away, what happens in terms of radiation to an object as it moves through this bubbly nothingness? He then calculated the answer to this question using the same quantum field theory in curved spacetime methods used by Hawking. Here, the "curved" spacetime isn't really curved at all. We are dealing with a subtle element of relativity. An object that is being accelerated seems from its own perspective to be in a curved or distorted region of spacetime because of an aspect of relativity called the principle of equivalence, which says that an accelerating observer acts like an observer at rest in a gravitational field. (In other words, you cannot tell the difference in effect between yourself accelerating or standing still surrounded by a gravity field.)

The result of Unruh's calculation is that the object absorbs radiation as though it were surrounded by radiation of a certain temperature, where the temperature in Unruh's formula is proportional to the acceleration of the object. At first sight, the Unruh effect seems paradoxical. There is no actual thermal radiation, or at least none that is visible to an ordinary non-accelerated observer. Why then does the accelerated observer detect thermal radiation? The answer to this paradox comes from doing a calculation of the Unruh effect in a second way, this time using the methods of ordinary quantum field theory.

Even in ordinary quantum field theory, as we noted, the vacuum is not empty but contains pairs of virtual particles being created and then annihilating each other. From time to time, the accelerated observer absorbs one particle of the pair, leaving the other particle of the pair to travel away as a real particle. In the language of ordinary quantum field theory, the accelerated observer is emitting particles through its interaction with the vacuum; but the way that the observer reacts to this interaction is the same way it would react to absorbing particles from thermal radiation whose temperature is given by Unruh's formula. In the end, both calculations agree on the effects on the accelerated object. They just use different language to describe these effects. The curved spacetime calculation says that the object detects a thermal bath of radiation. The flat spacetime calculation says that the accelerated observer emits radiation and "recoils" with each emission, and furthermore that the pattern and amount of these recoils are just those that the object would suffer if it were immersed in thermal radiation.

Now at this point you may want to yell, "But what's really going on?" In a sense we can't answer that; in another sense we've answered it twice. Both of the views, the flat and the curved, tell us what is happening and show us a way of looking at what is happening. Because we are dealing with things so far from the perceived universe, we can't give answers that fit into our normal framework of perceptions; we need to use models that go beyond perception in order to take in the universe around us. This is not the first or the last time we have shown the necessity of abandoning the comfort of the question above. Yet, strange as it sounds, people eventually come to accept as "really going on" things that only a few generations ago would have been regarded as absurd. Right now most people accept the idea that their bodies are grown from individual cells with information encoded in long chain molecules (DNA), whereas two hundred years ago, the nicest possible response to such an idea would have been "Huh?"

Back to the subject at hand. As with the Hawking effect, the temperatures involved in the Unruh effect are small. This makes detection difficult, though an analogous effect can be observed for circular motion of high-speed particles in particle accelerators. But even without detection, the Unruh effect is on solid theoretical ground since it can be calculated with ordinary quantum field theory, a well-tested theory. The fact that ordinary quantum field theory and curved spacetime quantum field theory yield the same answer for the Unruh effect then lends credibility to curved spacetime quantum field theory, which in turn lends credibility to the Hawking effect, which is calculated using that technique. We are left then with accepting the Hawking effect because it fits well with other pieces of theory that are themselves well grounded. One might think of it as being built on stilts. It touches the ground, but relying on it is wobbly.

Though the Hawking effect resolves a paradox involving thermodynamics, it creates a new one, usually referred to as the paradox of loss of information. Recall that it takes only two pieces of information, the mass and spin, to describe a black hole. However, the star that collapses to form the black hole is much more complicated and it takes much more information to describe it (the characteristics of every particle making it up).

Thus the process of black hole formation involves a loss of information. Similarly, when an object falls into a black hole, there is another loss of information. Only the mass and spin of the black hole change, but we need more information than that to describe the disappearing object. One can always say that the information is only "lost" to those of us outside of the black hole and that it is still contained in the black hole interior, which we ignore.

This cosmic act of under-the-rug sweeping unfortunately doesn't work because of the process of black hole evaporation. Recall that due to their temperature, black holes are giving off radiation and are therefore losing mass. Though at present the 3-kelvin radiation of the universe makes black holes net absorbers of radiation, the expansion of the universe will eventually bring this temperature below black hole temperatures and black holes will begin to lose energy to their environs. This loss of energy causes the black hole to shrink. Finally a black hole in such an environment will evaporate.

To understand why, remember that the temperature of a black hole goes up as mass goes down. This loss of mass leads to an increase in temperature, which makes the black hole lose mass even faster. Eventually, the entire mass of the black hole radiates away and the black hole disappears. How long does this take? For a solar mass black hole, the time for it to evaporate is a staggering 10^{64} years. This is not only huge compared to the time scales we are used to; it is even huge compared to the age of the universe, which is estimated as about 1.4×10^{10} years, a mere 14 billion. It might be that no black hole would evaporate in the entire life of the universe, but even so the potential for evaporation means that we have to account for where the information goes, because when the black hole disappears, the information for the star(s) that created it and every particle that fell into it have also vanished from the universe.

We can scrounge around the black hole looking for places where the information might be kept, but we have little to go on. For example, though the black hole emits a thermal spectrum of radiation, very little information is contained in this radiation. A thermal spectrum is completely described by its temperature (one piece of information for the whole spectrum). There aren't many other possible candidates for holding the information.

At this point the reader may be tempted to say, "Okay, so information is lost. So what? That happens to me every time my computer crashes." But in quantum mechanics, the preservation of information (in this context called unitarity) is a fundamental principle. For reasons too deeply involved in the quantum theory for us to go into here, a theory of quantum mechanics without unitarity is vastly different at its foundations and therefore its consequences from the present quantum theory. But a great deal of present quantum theory has been tested and found to work. Faced with this puzzle, one could theorize either that (a) quantum mechanics must be modified to be consistent with the information-loss properties of black hole evaporation, or (b) the details of the Hawking effect must be modified to be consistent with unitarity.

For decades a debate has gone on between the pro-Hawking proponents of alternative a and the anti-Hawking proponents of alternative b. Note that the anti-Hawking faction does not deny the Hawking effect; they simply think that the details of the black hole evaporation process need to be modified so that the emitted radiation is not precisely thermal but instead contains in some subtle "coded" way all the information about the black hole formation process. The debate was formalized in a bet that Hawking and Kip Thorne made with John Preskill. Kip Thorne has done much work on the astrophysical properties of black holes and on the properties of gravitational radiation. Though he and Hawking are on the same side on this bet, Thorne had a previous bet with Hawing on naked singularities that Thorne won (on what Hawking considers a technicality). John Preskill has done much work on particle physics and cosmology, but his present interests lie in quantum information and the theory of quantum computers.

Hawking and Thorne bet that information is lost in black hole evaporation and Preskill bet that it isn't. In the summer of 2004, Hawking switched sides in the debate and conceded his bet. However, though Hawking has given up, the pro-Hawking faction has not. At the time of the writing of this book, Thorne has not conceded the bet, while Unruh draws parallels with an earlier scientific controversy by saying, "Hawking recanted his theory, but then so did Galileo." The matter is still decidedly up in the air (actually much higher up than that).

We have brought this aspect of black hole study up to the present moment and presented one of its current controversies. It is tempting to look at the controversy and say that science can't handle it. ("O, woe is science. Doomed! Doomed! Doomed!") Not really. At any given point in time, such scientific controversies exist and are eventually resolved. That the answers do not yet exist tells us nothing about what will happen in the future. These questions in science lead to further theory and experiment. It is through unresolved questions like this that science grows because important questions motivate research in theory and in experimentation. The questions mark out the boundaries of the detected and theoretical universe as if to say: "Dig here!" *X* the unknown marks the spot.

So new scientists come along, dig up answers, but they also dig up new questions. In some respects, the current unknowns of science are a jobs program for the next generation of scientists, as each new stage of answers brings new questions and the need for more people to answer them.

CHAPTER 8

Gravitational Waves

Despite its tantalizing analysis of the characteristics of black holes, the Hawking effect is for now firmly theoretical, so we must ask again what we can, at present, detect about black holes. The observations of the effects of accretion disks around black holes seem to tell us little beyond the fact that a black hole is a compact object with a strong gravitational field. Is there anything else about black holes that we can detect and that will yield any more information about them than the simple fact of their strong gravity? There is something, and it has to do with a phenomenon known as gravitational radiation.

Gravitational radiation is analogous to electromagnetic radiation, something we are more familiar with. A radio station makes radio waves by driving an electric current rapidly up and down in its antenna. This makes electromagnetic waves that propagate outward from their source. When a radio wave comes to the antenna of a radio receiver, it makes electric currents move up and down in that antenna, and these currents are detected by the receiver. This, by the way, is a case where a very abstract aspect of the theoretical universe (Maxwell's equations of electromagnetism) leads to a relatively simple way to make objects: radio transmitters and receivers, which in turn lead to radical changes in human life (the entire interconnected web of communications we rely upon today).

Just as electric charges moving back and forth create electromagnetic radiation, so masses moving back and forth create a phenomenon called gravitational radiation. And just as electromagnetic radiation can be modeled as light waves, so gravitational radiation can be modeled as a kind of wave called a gravitational wave. These gravitational waves propagate outward from their source, and when they encounter other masses, these other masses are moved by the gravitational wave (this is exactly analogous to the radio transmitter and receiver). The main difference between these two kinds of radiation is that gravity is much weaker than electromagnetism, and this makes gravitational radiation much more difficult to detect than electromagnetic radiation.

That gravity is much weaker than electromagnetism is perhaps best illustrated by the example of a magnet lifting an iron nail. Here the magnet, a little object easily held in the hand, is pulling upward on the nail with its magnetic force, while at the same time the *entire Earth* is pulling down on the nail with its gravitational force, and yet the tiny magnet wins!

Though gravitational radiation is weak, it is not completely theoretical. It has been detected indirectly in a system called the binary pulsar. This system, first detected by Russell Hulse and Joseph Taylor in 1974, is a binary star system where both stars are neutron stars and one is a pulsar. For our purposes, an important property of this system is that the regularity of the pulsar's pulses make it an extraordinarily accurate clock, as good as (or perhaps better than) our best atomic clocks. But remember that the Doppler effect changes our observations of the pulse period depending on the speed of the pulsar. We can thus use the underlying regularity of the pulse as a means of measuring the speed. Because of the high accuracy of measurements of the pulse rate, we get very accurate measurements of the speed of the pulsar over time, and from that very accurate measurements of the orbits of the neutron stars, the masses of the neutron stars, and the orbital period (the time it takes the neutron stars to go in orbit around each other). Notice that just one very accurate measurement of a single phenomenon can be put through a number of calculations to produce accurate measurements of other related phenomena. The binary pulsar is a very good system, since it tells us how to detect and calculate its own characteristics. If only more things were like that. Of course, as with so many of these interesting star systems, you wouldn't want to live there. And you're better off touring them from a distance through telescopes.

Since the neutron stars are moving masses, general relativity predicts that the binary pulsar emits gravitational radiation. Furthermore, given the numbers for the neutron star masses and the details of the orbits, general relativity makes a precise prediction for how much energy the binary pulsar will lose in a given period of time, radiated away in the gravitational waves. This loss of energy should affect the orbits of the neutron stars. At first one might expect that a loss of energy would lead to the stars slowing down and therefore to the orbital period becoming longer. However, for a gravitationally bound system, a loss of energy means that the system is more tightly bound; in other words, that the neutron stars are closer together. Just as in the solar system, a smaller orbital size means a smaller orbital period (planets closer to the Sun orbit faster: they have shorter years than ones farther out).

So, as the neutron stars lose energy, they speed up. For our purposes, what is important is that general relativity makes a precise prediction for the (very small) change in the orbital period due to loss of energy in gravitational radiation, and that a pulsar is such an accurate clock that this small change can be (and has been) measured. The measured orbital period change agrees with the prediction, thus yielding an indirect detection of gravitational radiation. In other words, we can tell there is gravitational radiation because the system acts as if there is. If there were some other cause for the energy loss, then that other cause would have to produce exactly the same effects as the predicted gravitational radiation. That is certainly possible, but most unlikely.

Nonetheless, we would also like to have a direct detection of gravitational radiation. In analogy with neutrino detection, another situation where rarity of interaction made detection difficult, we can guess that the weakness of gravity will make it challenging to directly detect gravitational radiation. However, in analogy with successful neutrino detectors, even without knowing the mechanism of gravitational wave detection, we can guess that a successful detector must have three properties that we shall call bulk, precision, and insulation.

The analysis to follow points out a crucial method for getting along in the detected universe, that of analogous problem solving. If we have once solved a problem by a certain process and we encounter another problem with similar difficulties, we might well first try to solve it using a process similar to the method we used to solve the first problem. In short form, a good rule of thumb is not to reinvent the wheel, unless you're using square wheels.

Recall that in a neutrino detector, the more atoms present in the detector, the more likely it is that a neutrino would interact with an atom in the detector, so bulk in the sense of a large number of atoms was an essential feature. However, precision in the sense of being able to detect a single neutrino interaction was also essential. Finally, since other particles (for example, cosmic rays) could interact in the detector in ways that could be mistaken for neutrinos, it was necessary to insulate the detector from them by placing the detector deep underground.

Though, as we shall see, the mechanism for detection of gravitational radiation is factually very different from that for neutrinos, the same principles of bulk, precision, and insulation apply in its design and use. A gravitational wave detector is essentially a Michelson interferometer, the same apparatus used to show the constancy of the speed of light, though here the light source is a laser. The laser beam is split in two by the beam splitter, and each beam travels down one arm of an L and back. Then the beams recombine and interfere. For gravitational wave detection, the interferometer is set for complete destructive interference: the two beams cancel each other out, and in the absence of gravitational waves, the combined beam is completely dark.

What then happens to the interferometer when gravitational radiation passes through it? In relativity, gravity can be thought of as a distortion of space and time. This idea has been explained many times in books on relativity, using the image of a rubber sheet with masses put on it. The heavier the mass, the more it distorts the rubber sheet. The rubber sheet metaphor is okay as an image to communicate the basic idea of curved spacetime, but lacks much of the nuance needed to really get a grip on relativity.

Effect of Gravity Waves on Interferometer (Not to Scale)

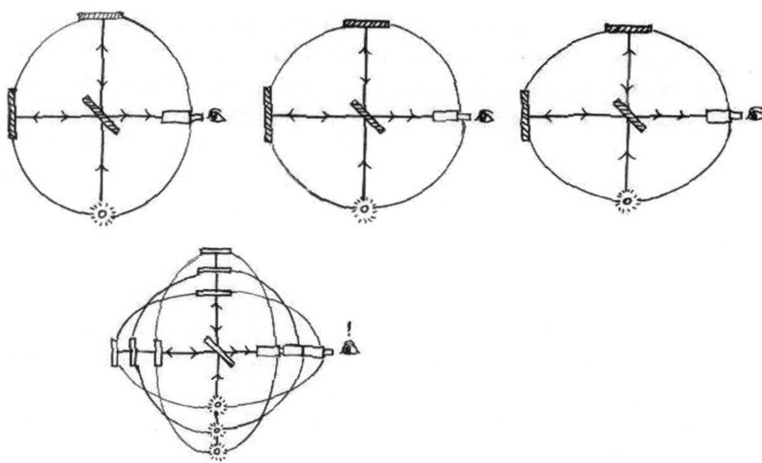

Figure 9

In any case, if gravity is bending spacetime, it is not surprising that for an interferometer set to destructive interference, the presence of a gravitational wave distorts space in such a way that the combined beam that was formerly completely dark becomes somewhat light. This happens because the wave shrinks one arm of the interferometer whenever it lengthens the other, and thus the interference is lessened (see fig. 9). From the amount that the beam lightens, we then find the strength of the gravitational wave.

From what has been said so far, you might be tempted to start building a gravitational radiation detector in your basement. (How hard can it be? Let me go to the store and pick up a laser, a couple of mirrors, and one of those beam-splitter gizmos.) But here is where the properties of bulk, precision, and insulation come in. The gravitational wave distorts an arm of the interferometer by changing its length by a certain very small fraction of the arm length, where the fractional length change depends on the strength of the gravitational wave. But since the signal in the interferometer depends on the actual length change, this means that the larger the interferometer, the larger the signal. So for gravitational wave detectors, the bulk that is needed is a large size. LIGO, a recently constructed gravitational wave detector, has an arm length of four kilometers (so much for setting it up at home). The odd-sounding name LIGO is an acronym for Laser Interferometer Gravitational-Wave Observatory. LIGO as a project actually refers to two such detectors, one in Livingston, Louisiana, and the other in Hanford, Washington. Other gravitational wave detectors have odd names like VIRGO, in Cascina, Italy; GEO600, in Hanover, Germany; and TAMA300, in Tokyo, Japan.

Even with this large arm length, the change in length of the arm is still a tiny fraction of the wavelength of the light. The detector must therefore have precision in its ability to accurately detect the tiny amount of light that comes through when the gravity changes the interfering laser beams from complete destructive interference to "not quite complete" destructive interference. Since the amount of light that comes through is also proportional to the power of the laser, the detector must thus have another kind of bulk in the sense of a powerful laser. (There goes buying the equipment at the hardware store.)

These properties of bulk and precision are, however, not enough on their own because things other than gravitational waves can cause a signal in the detector. The detector needs to be insulated from these outside influences. One simple example of insulation comes from the fact that the laser light can be distorted in its passage through the arms by the air that it goes through. Therefore, the entire L must be in pipes from which the air has been pumped out. LIGO must operate in a vacuum. The mirrors and beam splitter can be moved by vibration; so they must be mounted on elaborate suspensions that, to the extent feasible, insulate them from any vibration. As you may gather, we are definitely outside the realm of home experimentation.

Not mentioned in the above is the desire to have relatively strong sources of gravitational radiation. Just as the neutrino detector succeeds because the Sun pours out so many neutrinos, a gravitational wave detector needs what passes for strong gravitational waves. One source of such gravitational waves is the collision of black holes. Just as some binary star systems consist of a pair of neutron stars, so there are surely binary star systems that consist of a pair of black holes. Like the binary pulsar, a binary black hole system would emit gravitational radiation and the resulting loss of energy would bring the black holes ever closer together. Eventually the black holes would come close enough to collide and merge into a single larger black hole. The collision of the neutron stars in a binary neutron star system would also form a black hole and gravitational radiation, as would the collision in a binary system consisting of one black hole and one neutron star. This kind of collision is a tempting target since it would produce a strong gravitational radiation signal ("strong" being a very relative term).

If we can theoretically model what gravitational radiation from a binary black hole collision would be like, then we could use our detectors to try to find such a phenomenon. So what does theory tell us? For the final stage of the process at least (when the black holes merge), the prediction is clear. The final large black hole will settle down to a black hole described by Kerr's formula. Therefore, the process of settling down to this shape will consist of a distorted Kerr black hole shedding those distortions in the form of gravitational waves. It is helpful to think of this process as analogous to the ringing of a bell. A bell has a particular shape, but when the bell is rung, that shape is momentarily distorted. The bell vibrates, emitting sound waves, and eventually settles down to its undistorted shape. The character of the sound waves—in other words, the note of the bell—emitted in the settling-down process depends on the size and shape of the bell and on the material that it is made of. Analogous to a bell, the properties of the gravitational waves emitted by a distorted black hole in this "ring-down" process depend on the properties of black holes. But there are only two such properties: mass and spin. Hence the characteristics of the waves can be computed using Kerr's formula. If these waves are detected and match the prediction, this would verify Kerr's formula for the properties of black holes and give us detailed detected information on the properties of black holes.

But how often do black holes collide? Unfortunately, not very often. In a binary black hole system when the black holes are widely separated, the gravitational radiation is weak and it makes only a small change in the black hole orbit. It takes a long time for the black holes to get close enough to collide. The binary pulsar orbit is changing so slowly that it will take about 300 million years for the two neutron stars to collide. If we make the crude guess that binary black hole systems last this long, then in a year of observing time, only one out of 300 million of the binary black hole systems in the range of the detector will be observed colliding. The range of the detector is best explained by analogy with radio stations. The signal from a radio station gets weaker the farther you are from the transmitter, so there are only a limited number of radio stations that you can tune in. These are the stations that are in the range of your receiver in the sense that they are close enough so that their signal is strong enough to be detected by your radio receiver. Note that you can increase the number of radio stations that are in range simply by getting a better receiver.

At the time of the writing of this book, the gravitational wave detectors have not yet detected anything. This is not surprising given the scarcity of sources. Though LIGO could detect a binary black hole collision anywhere in our own galaxy or a nearby galaxy, estimates of the number of binary black hole systems make it highly unlikely that in a given year any such collision in our galactic neighborhood would take place. The solution is to get a better detector, one that can hear outside our galactic neighborhood. Though LIGO is an engineering marvel of bulk, precision, and insulation, it is not marvelous enough. Plans are in the works to upgrade the precision and insulation of the detectors, including a more sophisticated way to isolate the components from vibration. These upgrades should be in place within a few years of the writing of this book, and it is estimated that they will be sufficient to yield detections of black hole collisions.

An even more ambitious plan is to create a detector with vastly increased bulk by building a gravitational wave interferometer in space. Here the laser and the two mirrors would each be mounted on a separate satellite. These three satellites create an arm length much larger than LIGO (since the distance between each laser satellite and the mirror satellite would be the length of one arm). Furthermore, since the hardware is in space, the vacuum would be provided for free, as would removal of all Earth-moving and other terrestrial jiggling problems. This project is called LISA (Laser Interferometer Space Antenna). It is a planned NASA project, in collaboration with the European Space Agency, and may launch several years from now.

We started this step talking about black holes themselves as dark and difficult to detect compared to the Sun. We ended up with plans to try to measure their characteristics using their very hard-to-detect effects. In our next outward step, we will take up even darker matters and energies. However, if you will forgive the melodrama, from that darkness we will illuminate a clearer picture of our three-tiered universe.

Step 3

DARK MATTER —DARK ENERGY

Previous page: Dark matter. NASA, ESA, M. J. Jee and H. Ford et al. (Johns Hopkins University) 2007.

CHAPTER 9

Galactic Scale

The chapters to come are going to be very different from the preceding ones. Here we will deal not with science that is mostly known but with science that is mostly unknown. Bear with us for a moment while we pile up a little imagery.

Imagine a field of knowledge as being like an apartment building that is continually under construction. At the top of the building there is a swarm of workers constructing new stories, while in the completed parts of the building, the lower floors, people have already moved in and are living according to the ways the building has been built to work. They enjoy the amenities of the building, and they like or at least tolerate the style of decoration and the built-in appliances. Occasionally a worker comes down and remodels or fixes something in a previously built part of the structure, ripping out a wall, replacing an oven, repainting, and so on. Very rarely the building needs to be reworked on some fundamental level, rewiring, new plumbing, et cetera. When that happens, the crew drops back down for serious rehabbing and the apartment dwellers complain to the management, which shrugs and says, "That old Newtonian wiring couldn't handle the faster communication and heavier traffic. Einstein and Co. are putting in new cables from the ground up. Don't worry: you won't notice the difference until you need it."

In the buildings of science, most people live in the lower stories, using the results of science (technology and methods of problems solving) but not being at all concerned with how the building is being built on the top levels. People who need more of the science for the things they do in their lives (such as engineers) occupy higher constructed floors, but they do not need to know what is going on in the act of construction, nor must they worry about the infrastructure of the building itself, strange and interesting though that is. Anyone who has ever dealt with a building contractor will know that there's a lot of weird stuff hidden in the walls and floors of one's home. On occasion news trickles down from the top that some new floor is open for occupancy, and sometimes the noise of construction reaches the lower floors when an argument breaks out among the workers.

This separation between those who live in the building (the users of the field) and those who are building the building (the creators/discoverers of the field) is one of the things we have been trying to overcome in this book. We want to break down the barrier of mystery between living in the works of science and building the science itself. We do not expect or want everyone to become a worker in the fields of science, but we think things will be better if everyone understands more about what those weird noises are and why big dark holes have been cut in their walls.

All of which is by way of prefacing the point of this step: Welcome to the construction zone. Hard hats are required.

In the following chapters, we will look at elements of cosmology that at the time of this writing are the subject of ongoing research and discussion. Since we are dealing with the presently unknown, it may turn out that some of the ideas that we put in these chapters are wrong. This is not the case for most of the material of the previous chapters. We can be confident of what we wrote about the Sun, most of what we wrote about stellar evolution, and even much of what we wrote about black holes. In these chapters, much of what we say about cosmology, and about observations of dark matter and dark energy, is well established. But when we come to the nature of dark matter and dark energy, we enter the realm of the unknown, and confidence is not an asset in that realm.

Exploring the really unknown is where the methods of science fully show their value. When some new phenomenon is at first detected, the question "What is that thing?" begs to be answered. Theories will be created all over the place. Anybody can come up with something in the theoretical universe. Indeed, when a new thing is detected, it can be guaranteed that someone will posit the theory "It's nothing" and someone else will say some variation of "It is proof of my pet ideas." Theory alone, as we have said, is not enough. There needs to be detection to test the theory. Theory acts as a guide to creating the apparatus of detection, because if a thing is whatever is theorized, then there should be consequences of that and hopefully some of those consequences will be detectable.

We are going to lay out the paths presently being followed to create new science out of ignorance. The ignorance came about because something was found that could not at the time be well explained. Since then effort has been made to mine understanding from ignorance. We hope by showing these scientific works in progress to further illuminate the process by which the work of science is done.

To be fair, we could do this with any topic presently under new study. The ones we've picked, like the topics in the earlier chapters, have a certain dramatic quality: balls of fusing gas, galaxy-eating black holes, things like that. Dark matter and dark energy may have that level of drama—we don't know yet—but we do know that in trying to understand them, we have to reach to the beginning and the ends of the universe, so the search has its own dramatic quality.

In short, in pursuit of these topics, we hope to show some of the most interesting questions presently being asked about the composition of our universe. We will reach back to the beginning of time, hint about its end, conjoin the largest objects in existence with the smallest, and show how one blunder might hide two different truths.

Enough showmanship. On to the science.

✷

Typically, science endeavors to answer three basic questions about its subject matter: "What is that thing?" "How does it work?" and "What is it going to do next?"

We will begin, as we have before, with a seemingly simple question that is part of "What is that thing?": How much mass is there in a galaxy? This is a question cosmologists love to ask because mass is vital to most of the equations that govern the universe on large scales. For galaxies, one begins to answer "What is that thing?" by trying to measure its size and mass. Pictures taken with telescopes show the angular size of galaxies (how large they appear in the sky), which, once the distance to the galaxy has been measured, yields the actual size of the galaxy. We will postpone until later the tricky subject of how galactic distances are measured. As for galactic masses, recall that we find the masses of astrophysical objects by looking at their gravitational effects. This we will do in a bit. But we usually want to measure things in more than one way in order to make sure of our answers. Before we dive into the gravity, let us see if there is another way to measure the mass of a galaxy.

What we mostly get from astrophysical objects is light, and a galaxy gives us a lot of that. Is there a way to estimate the mass of a galaxy using the amount of light that we get from it? Sure. A galaxy is a collection of stars, and we have reason to believe that the Sun is a typical star. We know the mass of the Sun and its luminosity (the amount of light that it is putting out). If a galaxy has the luminosity of 1 billion suns, we might guess that it also has the mass of 1 billion suns. Put in the more technical language of astronomers, we know the mass-to-luminosity ratio of the Sun. We measure the luminosity of a galaxy and multiply by this ratio to get an estimate of the mass of the galaxy. This estimate of mass (which for our galaxy is about 30 billion solar masses) is just an estimate, since not all stars have the same mass-to-luminosity ratio and not everything in a galaxy is a star (for example, there's that big black hole in the center), but this number will do for a ballpark figure.

Here we are only considering "normal" galaxies, where most of the light comes from stars, rather than the active galaxies and QSOs treated in the previous chapters, where considerable light comes from an accretion disk around a central black hole. In a galaxy like that, the mass of the black hole and the light of the accretion disk do not correlate in a simple ratio.

Because of these difficulties, we want to measure the mass of a galaxy by using a more exact method. One such method is to detect the galaxy's gravitational effects and compare the measurement to the estimate using light. Since stars are held together in a galaxy by the galaxy's gravitational force, we should be able to find the mass of the galaxy using the motion of stars as they travel within the galaxy. Here we will use the same method that we used for the Sun, stars, and black holes: find the distance and speed of an object orbiting the galaxy and use these to calculate the mass of the galaxy. In other words, we want to measure the mass of the galaxy using the same old trick we used for every other big thing out there. As mentioned before, it is useful to see if one's old tools will work before building new ones.

We face one difficulty in using this tool. Stars are not orbiting around the galaxy, but instead are inside the galaxy. In this case, the same formula can still be used, but its meaning is a bit different. The quantity M in the formula for orbits that we used before is not the total mass of the galaxy, but instead the mass of that part of the galaxy that is at a smaller radius than the star's orbit. In effect, we pick a star and concentrate on the gravitational effect on that star of whatever in the galaxy is closer to the galactic center than that star. This process actually works, but it only gives us an estimate of the total mass closer to the center than that star is.

We can use this to measure the mass of almost the entire galaxy if we use as our measuring star a star close to the edge of the galaxy. Then we expect to get the best measurement of the total mass of the galaxy. We need, therefore, stars in the largest (that is, farthest out) orbits.

Let's try this measurement for spiral galaxies. Galaxies come in a variety of different shapes. Spirals are common, and we happen to live in a spiral galaxy, so we want to know about them first (we're parochial on an intergalactic scale). These galaxies, like our own Milky Way, consist of a round central bulge in the middle and a set of spiral arms in a disk around the central bulge. Our own Sun is in one of these spiral arms of the Milky Way.

The formula we are using to determine the mass of galaxies relates radial position (distance from the center) to speed. So instead of trying this with just one star, we can find the speeds and positions of many stars at different positions and make a graph of speed versus radius. Such a graph is called a rotation curve and should tell us something about the mass. As we noted in dealing with our solar system, orbital speed and orbital distance are connected to each other and the connection depends on the mass of the Sun. By analogy, orbital speed and distance in the galaxy would depend on the mass of the galaxy, so the graph should give us some idea of the distribution of mass within the galaxy and eventually point to the total mass.

Before doing the detections necessary to make such a plot, we can use theory to try to figure out what we expect the rotation curves of a spiral galaxy to look like. In a spiral galaxy, most of the light comes from the central bulge or regions near it. We would therefore expect that most of the mass is contained within the orbit of a typical star in the spiral arms. In other words, since we expect most of the mass to be in the center, we expect to get similar mass results for stars in the spiral arms, even if those stars are at different distances from the center. We would then expect the rotation curves to show smaller speed at a larger radius, just as we get for the planets in our solar system, where most of the mass is concentrated in the Sun. In other words, we theorize that a spiral galaxy is like a solar system, with the central bulge being like the Sun and the stars in the arms acting more like planets.

This is what astronomers expected when they did the work to make these galactic graphs; but it is not what they found when they actually did the plots (see fig. 10). As far back as the 1930s, the work of Dutch astronomer Jan Oort in measuring the motion of stars in the Milky Way had suggested more mass than could be accounted for by the stars. In the same decade, the maverick Swiss astrophysicist Fritz Zwicky was working on the motion of galaxies in galactic clusters and found similar results. At the time it was thought that the observations were not sufficiently accurate and galaxies not well enough understood, and that given time and progress in detection and theory, the puzzle would resolve itself in due course. Zwicky had the temerity to suggest that galaxies and clusters contained substantial amounts of dark matter (that is, matter that was not giving off light; as a term, "dark matter" is actually pretty dull, but it became interesting over time). However, Zwicky was well known as a man with numerous crazy ideas, and his dark matter idea was not widely accepted.

It was not until the work of Vera Rubin, starting in 1970, on the motion of gas clouds in many galaxies that the stark fact of dark matter became clear. Instead of the expected behavior in the outer regions of the spiral arms, the speed does not change very much as the radius increases. The rotation curve becomes "flat." This indicates a distribution of mass that is not concentrated in the

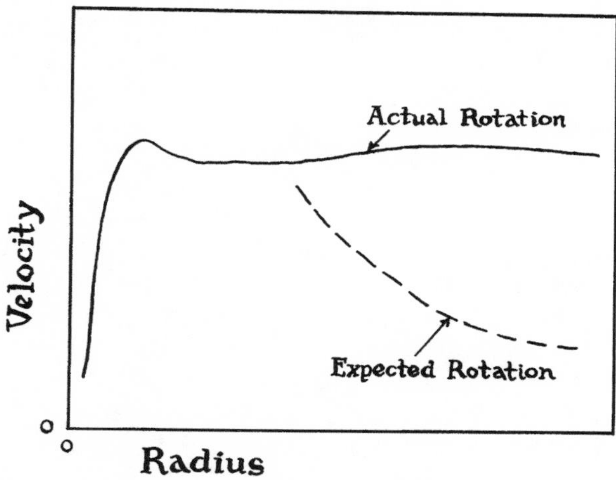

Figure 10

center. This phenomenon of flat galactic rotation curves puzzled astronomers: Why was there this other mass that did not follow the layout of the stars? Within the "What is that thing?" of the galaxy lay another "What is that thing?"—in this case, "What's all this stuff we can't see?"

CHAPTER 10

What Is Dark Matter?

If the stars orbit as if the mass of the galaxy is not concentrated in the center, then the most obvious interpretation is that there has to be other mass that is not seen and is more evenly distributed than the stars are. It is this matter that is making the rotation curve so flat. In other words, the hypothesis is that galaxies contain much mass that is not in stars, and that mass is not concentrated in the center. Even in the outer reaches of the spiral galaxies where stars are very sparse, there is still plenty of this matter.

The existence of this matter comes from theoretical interpretations of detected stellar motions. It is a perfectly reasonable interpretation: gravity depends on mass, too much gravity ergo more mass. But what to call this hypothetical stuff? Since the matter in question does not give off much if any light, it was dubbed "dark matter." Although a perfectly straightforward name, "dark matter" sounds all spooky and mysterious, which adds a strange cachet to this stuff. (The stuff is itself odd, as we shall see, but that doesn't justify the creepy overtones.)

We might well object to the rotation-curve-shape method as justification for believing in dark matter.

But even without looking at the curve shapes, just using the mass formula and detections, we arrive at a total galactic mass that is much higher than that obtained by using the light-based estimate above. Even if dark matter is not distributed in some weird way, there is darn good reason to accept that it exists and that there is a lot of it. The total mass of the galaxy as obtained by gravity methods is more than ten times the mass obtained by light methods. Put bluntly, the galaxies are mostly dark matter.

One can complain that we were being a little cavalier with the gravity method used above, since to arrive at our figure for the mass, we sat around playing tricks with the ideas of mass and orbit. But there is a way to find the mass of galaxies using a method closer to that found for measuring the mass of stars. Instead of looking at individual galaxies, we can use the method used for stars to find the mass of a cluster of galaxies. Just as stars are clumped into galaxies, galaxies are clumped into clusters. Just as the galaxies are held together by the gravity of star pulling on star, so the clusters are held together by galaxy pulling on galaxy. The universe has a clumped, somewhat curdled structure, with relatively dense and relatively empty areas. This lumpy appearance can be detected by direct observation with high-powered telescopes.

Each galaxy is moving within its cluster but kept from escaping by the gravity of the cluster. On the scale of clusters, galaxies are individual objects with their own motions and their own orbits. It may sound odd to say that a bunch of stars and empty space is on a certain scale one thing, but on the scale of planets, every individual object in your life is just part of Earth, and on the scale of humans, all the protons, neutrons, electrons, and empty space that make up the atoms of your body are just you.

If we see a galaxy as one thing, then the cluster is a relatively simple set of bodies in motion, just as a solar system is a relatively simple set of bodies in motion. The speed of the galaxy can be measured using the Doppler effect, just as we did for stars. Then by using the average speed of galaxies in the cluster as well as the size of the cluster, the mass of the cluster can be measured and compared to the mass estimate obtained by using the light method. Here we are on safer ground, using tried-and-tested detection methods without any game playing. But any hope that dark matter was an aberration of the means we used for measuring is quickly dashed by this method. As first shown by Zwicky in the 1930s, the same sort of result is obtained. The mass-to-light ratio of a cluster of galaxies is considerably larger than the mass-to-light ratio of the Sun—which means that galaxy clusters contain much matter that is dark.

What is surprising about these observations is how large the mass-to-light ratio is. It ranges from about ten times that of the light method for certain galaxies to a few hundred times that of the light method for some galactic clusters. That is, galaxies and clusters of galaxies contain much more matter than we would have guessed just from the crude estimate of mass using light.

Now that we know that there is dark matter, we come to the first question of science: What is it?

We shall adopt the strategy of the police captain in *Casablanca*. We will "round up the usual suspects." That is, we will consider, based on what we have covered so far, what matter a galaxy might contain that is considerably darker than the Sun. From the material of the last two steps, we would place the following on the list of usual suspects:

1. Gas clouds
2. Low-mass stars
3. White dwarfs
4. Neutron stars
5. Black holes

Let's pull these jokers in for questioning and see whether they have any alibis. Suspect 1: gas clouds. We know that gas clouds collapse to form stars. At the present time, some clouds have collapsed and some have not. In particular, for a cloud to collapse, it must be cold enough that its gravity overcomes its pressure. Thus we might guess that dark matter is simply gas clouds, especially those that are too hot to collapse.

Suspect 2: We know that stars come in a large range of masses and that those with larger masses have more gravity to overcome and therefore must burn their fuel at a faster rate in order to generate the pressure needed to balance gravity. But this also means that stars with mass much smaller than that of the Sun burn their fuel at a much slower rate and therefore put out much less light. Put another way, the smaller the mass of a star, the larger its mass-to-luminosity ratio. If low-mass stars are much more numerous than stars with mass close to that of the Sun, then low-mass stars might be dark matter.

Suspects 3–5: We know that stars do not burn fuel forever, and that eventually each star becomes one of three kinds of "dead stars"—white dwarf, neutron star, or black hole. If there are far more dead stars than live ones, this might explain the large mass-to-luminosity ratio of galaxies. Dead stars might be dark matter.

In the lineup, all of our suspects are pretty darn suspicious.

Which if any of these suspects might be the "guilty" one? Perhaps it isn't just one of them. Maybe they're working together, some sort of heavyweight gang operation, a dark conspiracy out to manipulate the universe with gravitic consequence. Sorry, we'll try to curb the melodrama . . . a little, and the puns, not much. Anyway a lineup isn't giving us what we want, so we'll let our suspects go, but we'll put them under surveillance and see what they do. To do that we need to detect their presence and see whether any of them are sufficiently numerous to be dark matter. (Remember that this is the work of a big gang. We warned you about the puns.)

Suspect 1: gas clouds. These crooks are pretty easy to spot, for though the clouds of gas in galaxies and clusters do not give off much visible light, they do give off light in other wavelengths. Those that are very hot give off light predominantly at wavelengths shorter than that of visible light, particularly X-rays. Those clouds that are very cold give off light of wavelengths longer than visible light. In particular, radio waves of wavelength 21 centimeters are given off by cold hydrogen atoms when the spin of the electron changes direction. Since the gas clouds are largely hydrogen, this gives us something to detect. Using X-ray telescopes, we can measure the amount of hot gas in galaxies and clusters, while radio astronomy (with the radio tuned to the "station" at 21 centimeters) can check for the presence of cold clouds of gas.

Those are good tests, and we can add another one. Remember that in addition to emitting light, gas can also absorb light. Since QSOs are very far from us, it often happens that the light from a QSO passes through a galaxy (or a cluster of galaxies) on its way to us. In this case, if the light of the QSO passes through a gas cloud, the spectrum of the QSO would show absorption lines from the gas cloud that it has passed through. (Remember that absorption lines are dark bands in a spectrum that lie at the same spots on the spectrum that the light bands from the elements would be.) An examination of the strength of these lines allows us to infer the amount of gas that the light has passed through. Thus, even though clouds of gas are comparatively dark, we can use light to detect their presence and measure how much mass they contain.

The result of this surveillance is a partial answer to the mystery of dark matter. Galaxies and clusters do indeed contain substantial amounts of matter in the form of clouds of gas. In the case of clusters, there is more matter in gas clouds than in stars. However, this is only a partial answer because the amount of matter in gas clouds is still much less than the amount of matter found from gravitational effects. We can charge gas clouds with misdemeanor dark mattering, but we haven't found the felon yet.

Unfortunately, the remaining suspects seem even less plausible than gas clouds since they are all variations of regular stars. Let's take suspects 2 and 3 together: low-mass stars and white dwarfs. We know how many stars there are with masses near that of the Sun. If low-mass stars were dark matter, there would have to be vastly more of them than of the moderate-mass stars. In addition, recall that in stellar evolution white dwarfs are the end states of the evolution of moderate-mass stars. If there are vast numbers of white dwarfs, there would have to have been vast numbers of moderate-mass stars earlier in the history of the galaxies.

Similarly, neutron stars and black holes are the end states of very massive stars and are the products of supernova explosions. If there are vast numbers of neutron stars and black holes, there would have to have been vast numbers of massive stars and supernova explosions earlier in the history of galaxies. None of this seems likely given current theories of star formation and of the evolution of galaxies. But this is only testing theory with theory. Since dark matter is a challenge to theory, we need to deal with it using more than theory. We need to detect and count the numbers of our suspects.

✹

There is an experimental technique that allows us to find the number of low-mass stars, white dwarfs, neutron stars, and black holes in a galaxy: gravitational microlensing. Using this method, we can see if there are enough of such suspects to be dark matter. In the last chapter, we examined black holes by considering gravitational fields so strong that they can prevent light from escaping, but it does not take a black hole to affect light gravitationally. Even a comparatively weak gravitational field has an effect on light. Each object, no matter how small, pulls light toward it just as it does all other masses in the universe. The idea that the path of light can be bent by a mass follows from general relativity: gravity is curved spacetime, and curved spacetime affects all things that travel in it, even light. Detection of the bending of starlight by the Sun was one of the first successful tests of general relativity.

Since ordinary lenses made of glass also bend light, this phenomenon of the bending of light in a gravitational field is called gravitational lensing. Just as an ordinary lens can concentrate light, so can a gravitational lens. In other words, in certain ways every object in the universe can be used like a lens (see fig. 11).

Objects whose light is distorted by a gravitational field may seem brighter than they would otherwise. Because a stronger gravitational field causes a larger effect, gravitational lensing is usually studied in the case where the lens is a galaxy or a cluster of galaxies, since they have a large enough mass to really bend the light of any object behind them. Because a star is much smaller than a galaxy, lensing by a single star is called gravitational microlensing.

If we want to look for suspects in our galaxy, we need to find a place to search for them where they won't be crowded out by the light from stars. The outer region of a galaxy, which is sparse in stars, is called the halo. Any of suspects 2–5 in the halo of a galaxy is called a Massive Compact Halo Object, or MACHO for short. (This deliberately cute acronym was chosen to contrast with the WIMPs to be discussed later.) Using appropriate apparatus, we can detect MACHOs using microlensing. If we found a lot of them in the halo, that would indicate there were a lot of them in the galaxy and might convict suspects 2–5 of flagrantly MACHO behavior with dark intent (we are going to beat this gag so far into the ground that it won't be able to get up until the end of time).

The process of finding MACHOs this way is a little tricky. Our galaxy is not alone. The Milky Way has a few much smaller "dwarf" galaxies in orbit around it. Suppose that we are using a telescope to observe a star in one of the orbiting dwarf galaxies. The line from the telescope to the star is called the line of sight. What happens if a MACHO in the Milky Way's halo passes near the line of sight? There is a point in the MACHO's journey that is closest to the line from our telescope to the star we are observing; this is called the point of closest approach. As the MACHO moves, it goes toward the point of closest approach, reaches it, and then moves on. The gravity of the MACHO distorts the light of the star

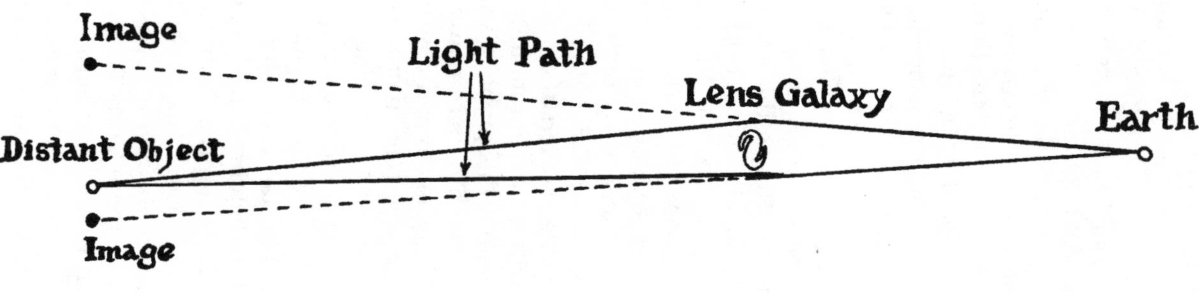
Gravitational Lens
Image
Light Path
Lens Galaxy
Earth
Distant Object
Image

Figure 11

and brightens the image in the telescope, with the largest amount of brightening taking place when the MACHO is at the point of closest approach. The brightening is caused by the gravitational lens focusing the light from the star toward our telescope. What we see in the telescope is a star that briefly brightens and then goes back to its normal brightness. When this temporary brightening happens, we conclude that some otherwise invisible gravity source served briefly as a lens. In other words, we are using gravitational microlensing not to find the thing we are observing (the star) but to find the lens (the MACHO).

This may sound silly, but the same kind of trick could be used with ordinary glass lenses. If a lens passes in front of your face, it will change what you see. If the thing you are looking at changes in the way it would if a glass lens passed in front of you, you can presume a lens did pass before you.

The effect of a MACHO on the observed brightness is very small, except when the distance of closest approach is itself very small. This happens when a MACHO almost crosses the line of sight. For any given MACHO in the halo and any given star in the orbiting galaxy, it is highly unlikely that we would observe

the MACHO lensing that star. We can overcome this problem by taking a lot of observations. Here the issue is similar to that of solar neutrino detection, where the probability of a given neutrino interacting with a given atom in the apparatus was small. For neutrino detection, this problem was solved by having an apparatus with a large number of atoms for the neutrinos to interact with. Similarly, microlensing searches involve observing a very large number of stars in an orbiting galaxy.

Neutrino detectors had to be placed underground to avoid the possibility that one of the much more numerous cosmic rays might be mistaken for a neutrino. Microlensing searches face a different problem of mistaken identity: variable stars. Some stars are variable—that is, their brightness is not constant but changes with time. Recall that a star is in hydrostatic equilibrium; gravity balances pressure. In a variable star, the hydrostatic equilibrium of the outer layers holds only in an average sense. At certain times the pressure in these layers is stronger than gravity, making the outer layers expand and causing the star to give off more light. This then cools the outer layers, so that pressure becomes weaker than gravity and the layer contracts. This is a cyclic process with the star repeatedly going through periods of brightening and dimming.

A microlensing search must have a way of telling a true microlensing event from a variable star. There are two helpful differences between these phenomena: (1) A lensing event affects all wavelengths of light equally, while when a variable star brightens, the light at some wavelengths brightens by a greater amount than the light at other wavelengths. (2) A variable star goes through a cycle of alternating brightening and dimming, while a microlensing event involves a single brightening and then a return to normal brightness. These two differences can be used to discard the variable star observations, leaving only the true microlensing events. As a by-product of this research, astronomers had to carefully observe and analyze variable stars, obtaining a wealth of useful information about them. So in order to separate suspects from nonsuspects, the nonsuspects had to be investigated carefully. Sometimes one line of research benefits another project in a different field, although it's not that great a method for police work.

These microlensing searches have been done and they have yielded a clear result: There are MACHOs, but in numbers far too small to be dark matter. Despite their MACHO posturing, suspects 2–5 are only minor criminals. We need to seek elsewhere to find the dark figures lurking in the interstellar underworld (okay, maybe we've pushed this joke a little too far).

✷

The usual suspects have been ruled out. If this were really a criminal investigation, we might pause to consider the possibility that no crime at all was committed. In the context of dark matter, we should consider the possibility that there is no dark matter and that we are simply misreading the evidence. How could that be? Well, the evidence for dark matter comes from its gravitational effects, and in particular from the mass formula obtained from Newton's formula for the gravitational force. If Newton's formula is wrong, then so is the mass formula, and by relying on the wrong formulas, we could be led to suspect the existence of dark matter when there is none. Perhaps instead of looking for dark matter, we should be looking for an alternative theory of gravity.

We've already been down this road. Einstein noticed that Newton's theory of gravity was not consistent with the principle of relativity, and he spent ten years looking for a better theory, finally coming up with general relativity. In science any search for a new theory must follow the principle that one shouldn't throw out the baby with the bathwater; that is, any new theory needs to agree with the old theory in those situations where the old theory has been tested. This was a stringent test for general relativity, which as we have seen it passed, by agreeing with Newtonian gravity in all situations where the gravitational field is weak. Any current alternative theory of gravity must pass an even more stringent test: It must not only give the same answer as Newton's theory for apples and planets; it must also agree with general relativity on such things as the small corrections to the predictions of Newtonian gravity for the orbit of Mercury, the bending of starlight by the Sun, and the change in the period of the binary pulsar.

Despite the inherent difficulty of such a task, there have been many proposals for alternative theories of gravity. (Or perhaps it's because of the inherent difficulty: some people like a challenge.) Generally, once careful calculations of the consequences of the alternative theory are done, it is found that the theory is ruled out by one or more of the existing experimental tests. Furthermore, an explanation of dark matter using an alternative theory of gravity must not only pass these tests but must also give the correct answers for the phenomena that we have so far interpreted as dark matter. Given the amount of data and the record of alternative theories of gravity, this does not seem very likely. We'll mark it down as a possibility that an alternative theory could clear everything of the crime. But this is sort of like waiting for a miracle to happen. We would have to sit around hoping that someone will create some as-yet-unknown theory that explains everything. This may happen, but we should exhaust all other possibilities before we come to rely on this hope.

We could at this point consider other suspects: lumps of rock, for example. There might be a lot more planets out there than we suspect. Planets are hard to detect, however (except the one we're standing on). And there might be other everyday stuff out there we can't easily find. But instead of trying to widen the field of suspects directly, we will use another experimental test that covers many of them.

For many crimes, DNA testing can exonerate anyone whose DNA does not match that found at the crime scene. Similarly, we will see that there is observational evidence that allows us to rule out not only the suspects that we have considered so far, but also a vast array of other possible suspects for dark matter. This test arises from the formation of chemical elements in the big bang.

We will now take a brief detour to the beginning of time in order to discuss the big bang and the formation of chemical elements and will then see how this applies to dark matter. This may sound like a weird thing to do, but consider our analogy with DNA testing. Genetics is one of the foundational aspects of biology, and DNA one of the root objects in genetics. Understanding the place of DNA in genetics permits genetic fingerprinting to be done; therefore, one of the roots of biology comes out in modern police work. So in modern astronomy and cosmology, we may need to reach back to the roots of the universe to test for something here and now.

The idea of the big bang came about through straightforward observation. Recall that we started this chapter looking at the properties of galaxies, wanting to know what they are: their size, shape, mass, distance from us, speed, and luminosity. Many of these observations were done in the 1920s at the Mount Wilson Observatory by Edwin Hubble. In college Hubble's main interests were astronomy and boxing. He gave up both to become a lawyer but quickly became bored with law and went back to astronomy. He was the first to show that there are other galaxies outside our own Milky Way and to measure their distances. The Hubble Space Telescope is named for him.

Hubble measured the distance and speed of numerous galaxies and found a curious fact: in general, the galaxies are moving away from us, and the farther from us they are, the faster they are receding from us. The mathematical expression of this fact is called Hubble's law, which can be written as the simple formula:

$$v = Hd$$

Here, d is the distance to a galaxy, v is the speed at which that galaxy is moving away from us, and the quantity H in the formula is called the Hubble parameter. Since both d and v are measured quantities, the value of the Hubble parameter can be found by using the formula and the results of those measurements. More precisely, given the measured distances and speeds of many galaxies, one can make a scatter plot of speed versus distance. Mathematically, Hubble's law says that the points of this scatter plot lie (more or less) on a straight line, while the Hubble parameter is the slope of that line. What this boils down to is that the farther the galaxy is from us, the faster it is moving away from us, double the distance, double the speed, and so on.

At first sight Hubble's law seems very strange. Why are all the galaxies running away from us? Is it something we said? More seriously, perhaps the Milky Way has the cosmological equivalent of body odor. To put it somewhat differently and more helpfully, how does it happen that we are living on precisely that galaxy that all the other galaxies are moving away from?

This question disappears, as does the apparent strangeness, when Hubble's law is examined more closely with a less parochial view. If we observe some galaxy X moving away from us at speed v, then any observers on galaxy X would see *us* moving away from *them* at speed v. More generally, an observer on galaxy X measuring the speeds and distances of galaxies relative to galaxy X would obtain the same Hubble's law. All the galaxies would seem to be moving away from galaxy X according to the same Hubble formula and with the same value of the Hubble parameter.

It is perhaps best to describe this phenomenon without using the point of view of any particular galaxy. We should give up our "Milky Way–centric" way of describing the Hubble phenomenon (which might be offensive to the equally parochial inhabitants of galaxy X) in favor of a more universal description. We do need such a description, because Hubble's law from this perspective says that every galaxy is moving away from every other galaxy and the speed at which they flee from each other increases as the distance between them increases. Hubble's law is talking about the way the universe is changing over time, not just the way any particular patch of it is changing.

To get this more universal description, note that since the galaxies are moving away from us, the distance between us and them grows greater as time goes on. Thus all observers, regardless of their galaxy of national origin, can agree that the distance between galaxies is getting larger. This is usually stated in the shorthand phrase "the universe is expanding." This does not mean that every object in the universe is getting bigger (or as Woody Allen put it, "We live in Brooklyn. Brooklyn is not expanding!"). It is simply that on a very large scale, space is expanding in the sense that the galaxies keep growing farther apart from each other.

How fast are the galaxies moving? If we can figure out their speed and distance, we can calculate the value of the Hubble parameter. This question is answered by measuring the speeds and distances of galaxies. The measurement of speed is fairly straightforward, using the same Doppler-effect technique mentioned previously. However, the measurement of the distances of galaxies is quite difficult, as we will explain later in the chapter, and it is only in recent years that a reasonably accurate value of the Hubble parameter has been obtained. The Hubble parameter H is approximately 70 kilometers per second per megaparsec. In other words, a galaxy that is 1 million parsecs away from us is receding from us at a speed of about 70 kilometers per second.

The units in which the Hubble parameter is expressed, kilometers per second per megaparsec, seem strange in that they do not apparently express any coherent concept. Most units like those for mass, distance, acceleration, temperature, or energy can at least connect to our mental idea of what thing in the world the unit is supposed to represent. But sometimes you come up with a quantity or idea you need to use that just doesn't fit your thinking. In those circumstances, you can either live with it or try to change your thinking to fit it. In this case, let's look a little more deeply and find out what changes in thought we need to make.

The units of the Hubble parameter are useful for astronomy because a megaparsec is a reasonable unit for intergalactic distances. A little dimensional analysis leads us to look at Hubble's law in a slightly different way. Kilometers and megaparsecs are units of length, so the units of the Hubble parameter are (distance/time)/distance. Distance divided by time and then divided by distance is the same as distance divided by distance divided by time. Distance divided by distance is a dimensionless number (one that has no units). So the units of the Hubble parameter turn out to be 1/time. Since 1 divided by time is a weird unit, we can look not at H itself but at $1/H$, which has units of time, and time is something we are familiar with.

If $1/H$ is an amount of time, how long a time is it? As a parsec is about 3.1×10^{13} kilometers, we find that $1/H$ is about 4.4×10^{17} seconds, which comes out to about 14 billion years. What is the meaning of this time? A quick examination of Hubble's law shows us that this time is a crude estimate of the age of the universe. If a galaxy is a distance d away from us, then according to Hubble's law, it is moving at a speed $v = Hd$. In the time $1/H$ that galaxy has moved a total distance from its starting point of $v(1/H) = Hd(1/H) = d$. But this is precisely the distance it is from us now. In other words, at a time $1/H$ ago, that galaxy and our galaxy were in the same place.

Applying this reasoning to all galaxies, we find that at a time $1/H$ ago all the galaxies were in the same place. Or, to be more clear, at that time everything that eventually becomes part of a galaxy was in one place, since we have no cause to believe that at that distant past time the stars that make up the galaxies had formed. Indeed, if we consider that at that time all the mass of the universe was in one place, we might well expect a very different kind of object from the mostly empty space we see in our present-day universe.

This much denser state where all the galaxies were in the same place is called the big bang, thus $1/H$ is the time since the big bang. That is, $1/H$ is the age of the universe. This is, as we said, a crude estimate because it makes the unwarranted assumption that at earlier times the universe was expanding at the same rate it is today. If at earlier times the universe was expanding faster, then the universe is younger than $1/H$, while if the universe was expanding more slowly at earlier times, then the universe is older than $1/H$. In the next chapter, we will consider how to do better than this estimate of the age of the universe.

When gas expands, it tends to cool off. If the universe has been expanding for billions of years, then it has been cooling off for billions of years. The early universe must have been much hotter and denser than the present universe. The universe started out very hot and very dense and rapidly expanded in all directions. In other words, the universe as we see it resembles the aftermath of a gigantic explosion: a "big bang." Incidentally, the name "big bang" was coined in the 1940s by Sir Fred Hoyle, one of the opponents of this theory, in order to ridicule the people he was arguing with, chiefly George Gamow, the main proponent of the idea of an explosive beginning to the universe.

In one respect, however, this analogy with an explosion can be misleading. An ordinary explosion starts at a particular place and expands outward from there. So it is natural to ask, "If the universe is expanding, what is 'outside' the universe that it is expanding into?" But there need not be anything outside the universe. The whole universe can expand (in the sense that distances between things continually get larger) without there being any such thing as an "outside" for it to expand into.

This may sound weird, but it can be analogized simply. Take an ordinary balloon that has not been blown up, and with a marker put a bunch of dots on the balloon. These dots will be fairly close together to begin with. Now blow up the balloon to full size and look where the dots are in relation to each other. They have all moved apart from each other, just as the galaxies have moved from each other since the big bang. In this metaphor the balloon itself is space and the dots are the galaxies. It is not just the galaxies that have moved. Space itself has gotten bigger.

But the immediate complaint can be lodged, "Hey, I'm out here blowing up the balloon, and I'm in this space. And besides the balloon's expanding in the space I'm in." Regardless of that part of the analogy, there need not really be an outside space for the balloon to blow up in. This is one of those matters that are mathematically simple but hard to communicate clearly because the math conflicts with people's ideas of what makes sense. There is a commonsense, intuitive idea of space and of objects being in space: the concept that everything is somehow inside some big, all-containing, unchanging container. This idea is comfortable for everyday thought. But this concept makes the idea that an object (or space) can change shape, expand, or contract without some space around that object for it grow or shrink into an uncomfortable one.

We could put in various ideas and images to make it easier to accept this kind of concept, and we will put one in later. But we would rather start with a general principle that is useful in science and beyond science. This principle was first formulated by William of Occam over six hundred years ago and is called Occam's razor (which is unfair, since Occam is a place—it should be William's razor). Occam's razor is often rendered into this idea: "In any situation the simplest explanation is likely to be the true one." This is not Occam's razor and isn't true. The real principle is "Do not create theoretical objects *unnecessarily*. Use an explanation that does not require extra objects or beings."

Let us take the case of where do Christmas presents come from? A child equipped with Occam's razor might reasons as follows: "My parents go out to various stores and come back with boxes and bags, or they place orders over the phone and the computer and then packages are delivered. It would seem that they are the ones getting the presents." Without Occam's razor, they would be more likely to accept the comforting (?) idea that an old man sneaks down their chimney and sticks the presents under a tree and then sneaks back up to the roof and magic flying reindeer whisk him away to the North Pole, where a squad of elves will make next year's toys. That's a lot of entities for something with a much simpler albeit less brightly colored explanation.

This actually illustrates the tension between Occam's razor and other kinds of explanations. Occam's razor talks only about necessity of entities. Necessity is not the same thing as comfort or a good story. In other words, just because we are uncomfortable with an idea does not justify adding more ideas that comfort us. We need to be able to accept the uncomfortable if it fits the facts better than the comfortable.

In the example of the expanding universe and the possibility of a space into which our space expands, we should look for that hypothetical space outside our space only if some phenomenon in our universe points to its existence (for example, objects suddenly appearing and disappearing as if they were passing through our space from elsewhere). There are no such phenomena that necessitate the theoretical creation of such a space.

The expansion of the universe does not directly point to the existence of some hyperspace—the mathematical and science fiction term—around our space. There may be one, but we have no reason to add it to our theories. The expanding universe on its own fits together, and it is we who need to change our thinking to go along with it. That seems fair—our minds are much more adaptable than the laws of physics. Ideas can change by learning and thinking; gravity just is. The universe as it is, is beautiful and harmonious in its own way, not in the way we would like it to be.

If after all that comfort is still wanted, we will offer the following uncomfortable comfort in the matter of the expanding universe. First remember that what we can observe is not the whole universe, but instead that piece of the universe that is within our past light cone: the observable universe. One can note that the farthest out one can see is the farthest distance that light could have traveled from the big bang until now. One can then take the point of view that we don't know for sure what is going on outside of the observable universe, but that in any case it is an empirical fact that the observable universe is expanding whether there is or isn't anything to expand into. In other words, there might be some outside space, but we can't detect it, so why worry about it.

*

The big bang tells us that the early universe was different from the present universe, lacking the distribution of galaxies, and indeed having all the mass of the universe in one place. We inevitably want to know what it was like. We already know that it was hot and that hot objects give off light, so the early universe must have contained light as well as matter. What has happened to that light? It is still here, but the light has cooled off as the universe expanded. That may sound weird—how can light cool down? Temperature is an average of the energy of the particles in a system (temperature of an object is the average kinetic energy of the atoms in that object). In the case of light, temperature is just the average of the energy of the photons. High-energy light is hotter; low-energy light is cooler. When light cools down, it redshifts into the lower energy spectrum (remember the energy of light determines its color). At the present time, that light has a temperature of about 2.7 kelvin, just slightly above absolute zero. Light at that temperature is strongest at microwave wavelengths. The afterglow of the big bang explosion is thus an ever-present bath of cosmic microwave background (or CMB in the usual acronym of astronomers for this phenomenon).

The CMB was found by accident by Arno Penzias and Robert Wilson in 1965. They were testing out a microwave antenna and noticed a mysterious source of microwave "noise." The noise did not seem to be coming from any particular direction, and they could not get rid of it no matter what they did. Eventually they realized that it was not a fault in their equipment but a fundamental part of the universe.

This notion of an accidental discovery—which we have already seen in the case of pulsars—may seem strange. We are used to thinking of science as purposeful, rational, methodical activity extending carefully down from theory to fact and up from fact to theory. But it is not at all strange when looked at from the point of view of the detected universe and the nature of tools. We create tools for specific purposes, but the tools themselves are wholly indifferent to those purposes. They work according to what they are, not what we want them to be. A simple example is the knife. We may create a knife for chopping vegetables, but someone may get the clever idea of using it to carve wood, or to stab someone. The knife does not care and will not refuse to act against our purpose for it. The same applies to detection equipment. Just because we call something a magnifying glass doesn't mean we can't use it to start fires or line it up with another lens to make things look smaller. Similarly, we may say we want to detect microwave signals from a satellite, as Penzias and Wilson did, but microwaves are microwaves and we may find a whole other source of them and so by accident discover the CMB.

From this perspective the CMB was always out there waiting to be discovered by the first scientist with a sufficiently sensitive detector in the right part of the microwave band. Because we are always making better and more varied detecting apparatus, ever more things can be detected. One might say that the detected universe itself is always expanding. Alternatively, one could say that astronomers are the ultimate followers of the advice given in the musical *Mame* to "open a new window."

Since the discovery of Penzias and Wilson, the CMB has been measured in great detail and with extreme precision at several different microwave wavelengths and at all directions in the sky. From these measurements, two remarkable features stand out. The first is that the microwaves are just those that would have been given off by an object at 2.7 kelvin, even though the light has not been in contact with any object in billions of years. The second is that that temperature is almost exactly the same no matter which direction of the sky one observes. Put together, these two observations mean that when the CMB light was last in contact with matter, that matter had almost exactly the same temperature and almost exactly the same density at all places, and so did the light. This is a very different state from the mostly empty space universe of today, so how did we get from there to here?

The measurements of the CMB were done in several different experiments and by many people. However, from this group, two experiments stand out as pointing in interesting directions (while looking all around): COBE and WMAP. COBE, an acronym for COsmic Background Explorer, was launched in 1989 and was the first experiment to show that the temperature of the CMB is not exactly uniform and to find the small variations in temperature with direction on the sky. These tiny variations are important, because a universe that started completely uniform would remain so forever: there would be no stars, galaxies, planets, or life. However, tiny non-uniformities at the time of the big bang grow over time under the influence of gravity, eventually forming the galaxies and stars that we see today. WMAP was originally named MAP, an acronym for Microwave Anisotropy Probe. Here "anisotropy" is a technical term for the fact that as one looks in different directions, one sees slightly different temperatures. MAP was launched in 2001. One of the leaders of the MAP team was the American astronomer David Wilkinson, who had spent his entire career studying the CMB. In 1965 Wilkinson was working on a project, led by Robert Dicke, that was attempting to find the microwave radiation predicted by the big bang theory, and that was scooped by Penzias and Wilson's accidental discovery. When Wilkinson died (in 2002), his collaborators added the W for Wilkinson to the experiment's name. WMAP has made stunningly accurate measurements of the CMB, which for the first time allow a new era of "precision cosmology." It is now possible to state with reasonable accuracy the age of the universe and the amount of matter that it contains.

The early history of the universe is essentially the history of a hot mass of light and gas expanding and cooling. As it cooled down, various processes became possible that were not so before because it had been too hot, and therefore various kinds of objects could come into existence that could not do so in the hotter early seconds of existence.

In earlier chapters we talked about how certain things (such as individual atoms) could not exist if there was too much heat. The earliest minutes of the universe consisted of rapid cooling that brought about the emergence of many different kinds of things at different temperatures. We wish to focus on a particular time, about three minutes after the big bang, when one process, that of the formation of atomic nuclei, became possible. This piece of history will lead us in a roundabout way to the test to eliminate candidates for dark matter. (Okay, it's been a long digression, but 14 billion years in a few pages isn't bad.)

Since matter is made of nuclei and electrons, and since nuclei are made of protons and neutrons, one way to pose the question "How much matter is there?" is to ask "How many protons and neutrons are there?" In order not to have to keep using the clumsy phrase "protons and neutrons" over and over, we will again adopt the standard particle physics usage and refer to protons and neutrons collectively as nucleons.

We will ask the "How many nucleons are there?" question in a funny way: using the Greek letter eta (η) as the symbol for the ratio of nucleons to photons. That is, count the number of nucleons in the universe, then count the number of photons, and divide the first by the second and that gives us eta. This may sound like a ridiculous idea, to count everything of a certain kind in the universe, but we can be sneaky about it. From measurements of the CMB, we can figure out how many photons there are (since they were all there in the CMB), so if we can find eta some other way, we can find out how many nucleons there are. But eta hardly changes at all between the time of nucleosynthesis (the formation of atomic nuclei) and now. This is because between that time and now, no new nucleons and comparatively few new photons were formed. So if we can find eta at the time of nucleosynthesis, we can find the amount of matter (or at least matter made of protons and neutrons) now.

Recall that in most layers of the Sun conditions are too hot for actual atoms to exist. Any atom there would lose its electrons due to high-speed collisions. Similarly, at sufficiently early times, the universe was so hot that no nuclei could form; any nuclei there would be blasted apart into their constituent protons and neutrons by high-speed collisions. The universe then consisted of protons and neutrons, as well as photons, electrons, positrons, neutrinos, and antineutrinos.

At first, the presence of neutrons in this mix seems surprising, since as we noted in an earlier chapter, a neutron left to itself will decay into a proton, an electron, and an antineutrino. However, the half-life of a free neutron (the time by which half of the neutrons decay) is about ten minutes, whereas the temperature of the universe became low enough for protons and neutrons to combine and form deuterium roughly three minutes after the big bang. That's plenty of time to snare some neutrons.

We can work out, given the charted development of the early universe, that existence started out with equal numbers of protons and neutrons. But three minutes later, there was only one neutron for every seven protons. This is only partly due to neutron decay. A larger role is played by reactions involving neutrinos and electrons, such as a neutron combining with a neutrino to produce a proton and an electron. This kind of weak-force reaction cut down the neutron population well before nuclei could form. However, once the temperature was cold enough for deuterium to form, nearly all the remaining neutrons combined with protons to form deuterium. This deuterium eventually formed helium, since while the universe was cooler than when it started, it was still easily hot enough for fusion to occur.

At the time of nucleosynthesis, there was one neutron for every seven protons. This means that one out of every eight nucleons was a neutron. Since that neutron then combines with a proton to form deuterium, this results in one-quarter of the nucleons (one neutron and one proton out of eight nucleons) being incorporated in deuterium. Since that deuterium eventually ends up as helium, this means that one-quarter of the nucleons end up in helium. Thus, a few minutes after the big bang, the composition by mass of the nuclei in the universe was about three-quarters hydrogen (the protons that did not find a neutron to combine with) and one-quarter helium (all the nucleons that got stuck together). These calculations rely only on detected reactions that still can happen today under the right circumstances. In other words, we do not need to hypothesize anything stranger than the big bang itself to come to these conclusions.

Remarkably now, billions of years later, the composition of the universe is not very different. It is still about three-quarters hydrogen and one-quarter helium. But now there is about 1 percent of heavier nuclei. Billions of years of nucleosynthesis in stars has done less to shape the composition of the nuclei in the universe than three minutes of big bang nucleosynthesis.

Though most of the deuterium was incorporated into helium, not all of it was. A small fraction of the deuterium produced in the big bang remains as deuterium. There also remained a small amount of helium 3 (a light form of helium composed of two protons and one neutron) and another type of nucleus called lithium 7, which is composed of three protons and four neutrons.

After these first few minutes of the universe, we had a lot of the nucleons in helium 4 (ordinary helium), a little in lithium 7, some in helium 3, and some in deuterium. The next question is a little weird and is asked not for the sake of the answer but because of the use we will make of that answer in answering another question. We ask (for reasons of mathematical sneakiness): What is the fraction of nuclei in the universe, by mass, in deuterium, helium 3, and lithium 7? It can be determined that this value depends on eta. At the time of big bang nucleosynthesis, there were far more photons than nucleons, making eta a very small quantity, about half a billionth. This means that there were about 2 billion photons for every nucleon. Hence more of the universe at the time was light (radiation) rather than nucleons (matter).

A larger eta would mean more nucleons, which means more nuclei, which means that nuclear reactions are more likely to take place and sets of nuclear reactions are more likely to proceed to completion. Deuterium and helium 3 are intermediate products of the nuclear reactions that lead to helium 4, so a larger eta would mean smaller amounts of deuterium and helium 3. This may sound peculiar, but if we consider deuterium and helium 3 to be ingredients in helium 4, and eta as a measure of how likely the process of making helium 4 is to be completed (because the ingredients, being more abundant, are more likely to collide with each other and thus combine), then a higher value for eta means more completion, hence less ingredients left lying around. The connection between eta and the abundance of lithium 7 is more complicated but still predictable, so that if we know eta at the time of nucleosynthesis, we can compute the abundances of deuterium, helium 3, and lithium 7 in the universe, and conversely if we know those abundances, we can figure out eta.

This sort of comparison is rather tricky, because both stars and the big bang produce elements through nucleosynthesis, so later processes than the big bang will affect the amounts of these elements. To untangle these two effects, a combination of both detection and theory is needed. On the detection side, one looks in places where the matter has undergone the least amount of processing in stars—this should give us regions that are more post-big-bang-like than those areas that are near to stars. Relevant methods include an observation of the element abundances in meteorites, as well as the detection of atoms in the interstellar medium (the extremely rarefied gas and dust between the stars) by examination of their absorption of light. On the theoretical side, one carefully calculates and tries to take into account the amount of nucleosynthesis in stars. For example, any deuterium produced in the big bang that undergoes further nucleosynthesis in a star is usually turned into helium 3. This process changes the amount of deuterium and the amount of helium 3, but not the amount of deuterium plus helium 3, so if we worry not about the individual values and concentrate on the sum of them, we can partially discount the stellar process. This combination of detection and theory allows us to find eta at the time of big bang nucleosynthesis, which is the same as eta now, and this in turn allows us to find the amount of matter now.

All of that sounds messy, and it is, but the purpose is a simple one. Eta is the ratio of nucleons to photons. We want to find the number of nucleons, so if we can find the number of photons (and we can from the CMB) and find eta (this is the messy part using detection of elements and theories of formation), we can find the number of nucleons, which is the amount of matter.

Or is it? What we can really find is the amount of matter made of protons and neutrons (and electrons). Cosmologists call this baryonic matter (after baryons, a collective name for a class of particles that includes protons and neutrons). All of our suspects—even neutron stars and black holes, since they come from the collapse of baryonic matter—fit into this category. We know how much baryonic matter there is, and we know the total amount of matter there is from its gravitational effects in the universe. Comparing these two quantities shows a startling result.

There is about seven times as much matter as there is baryonic matter. This is one of the most amazing and bizarre scientific discoveries of our times. The entire known contents of the universe, all the normal matter, has been cleared of the charge of being dark matter. Dark matter is something completely different from anything we are used to.

To which one might object, "Hey, wait, what about those neutrinos? There's a lot of them and they aren't baryonic. Besides, they're hard to detect. They sound like dark matter to me. Pull them in and see if they fit our criminal profile."

Calculations of the properties of the big bang show that in addition to the CMB photon background, there should also be a background of neutrinos at a temperature slightly lower than the CMB and with the number of background neutrinos close to the number of CMB photons. Since there are about a billion photons for every baryon, we would expect about a billion neutrinos for every baryon. Thus even if the neutrino has a tiny mass of about ten-billionths of the proton mass, the huge number of neutrinos would yield a total mass in neutrinos ten times the mass in ordinary matter. They would be our needed dark matter.

Furthermore (says the prosecution, building its case), neutrinos interact only through gravity and the weak interaction. Since the neutrino background is so cold, these neutrinos have very little energy. For the weak force, the tiny probability of interaction gets even smaller if the energy of the particle is small. (Remember, our neutrino detectors detected high-energy neutrinos coming out of the Sun. These background neutrinos are a lot colder.) This means that the only way that we would be aware of the neutrino background is through whatever gravitational effects it might have through its mass.

Aha! Now we've got you, you neutrinos! We accuse you of being dark matter.

Sadly, neutrinos have an alibi. And their alibi is huge. Every galaxy in the universe testifies to their innocence.

✸

The early universe, as we mentioned, was very uniform, with the density almost exactly the same at every point. The present universe is anything but uniform. Most of it is space that is almost empty, and what matter there is comes clumped in stars and gas clouds, which are clumped in galaxies, which are clumped in clusters of galaxies. How did the universe get from its smooth beginning to its present clumped state?

Gravity. Though the early universe was highly uniform, it was not perfectly so. Some regions had a slightly higher density than the average, some a lower density. But since matter exerts a gravitational force, this means that the slightly over-dense regions exerted a slightly higher gravitational force on their surroundings, pulling more matter in. Thus, over time, a mildly over-dense region became more and more dense, until finally these over-dense regions became dense enough for the processes of gravitational collapse that we previously discussed to take hold. The densest parts of these denser parts collapsed into the stars, galaxies, and galactic clusters that we see today. Exactly the reverse happens to an initially slightly under-dense region. It loses more and more matter to its surroundings until eventually it becomes a "void," a huge region of intergalactic space containing very few galaxies.

There are some very detailed, beautiful computer simulations of this process that can be found and played, but this is a book so we have to leave it to your imaginations aided by a little imagery (we won't bother to argue about the relative processing power and graphic advantages of both systems). Envision a cloud of gas thicker in parts, thinner in others. The thicker parts pull more and more gas into them. But within the thick regions there are themselves thicker and thinner neighborhoods. The gas collects into the thickest areas within the thicker regions within the thick neighborhoods, until deep enough into this process the really dense regions of gas begin to fuse, and the thickest parts cease to be clouds and become stars within empty space. The stars are themselves clustered in the remains of thick regions that are galaxies and the galaxies within thick regions that are galactic clusters and so on. Certain people have likened this process to the curdling of milk into cheese, so perhaps we should rename the Milky Way the Cheesy Way.

The present structure of galaxies, clusters, and voids can be observed with telescopes. The initial pattern of slight over-densities and under-densities can be inferred from observations of the slight variations in temperature of the CMB from one direction in the sky to another. Given the early universe over-densities and under-densities that are inferred from the CMB, and *if* astronomers knew the properties of dark matter, they could use the law of gravity (and a good-size supercomputer) to find the large-scale structure of galaxies, clusters, and voids that make up the present universe.

We don't know the properties of dark matter, but the idea of such a simulation can be turned around in the following way: Start with a guess as to the properties of dark matter and then make a simulation of the universe using those properties. Astronomers used this guess and the detected early universe over-densities and under-densities, and plugged them into the computer simulation of the formation of the large-scale structure of the universe (that is, the pattern of galaxies, clusters, and voids). If what resulted from the computer simulation did not resemble the large-scale structure that we see in the universe, then the dark matter model, the guess as to the properties of dark matter, would be ruled out.

Notice that this process is a sophisticated form of trial and error. An idea is formulated; its consequences are calculated. These are then tested against the detected universe. If it fails, the idea is discarded and another guess created. If it fails by a little, the original idea might be tweaked instead of thrown out completely. Using this process, it is possible not just to test particular ideas of dark matter, but to test what properties of dark matter (mass, temperature, speed, charge, and so on) contribute to the way in which the present universe formed.

For the purposes of large-scale structure formation, trial and error has determined that only one property of dark matter is important: the speed of dark matter particles at one specific time during the process of structure formation. Dark matter that was moving slowly compared to the speed of light at that time is called cold dark matter, while dark matter that was moving at an appreciable fraction of the speed of light at that time is called hot dark matter. These names, "hot" and "cold," are used because for a given material the hotter it is, the faster its particles move. It turns out that neutrinos are definitely hot dark matter. Their small mass ensures that at the time relevant for structure formation they were moving very rapidly.

Since the only thing that matters for the computer simulation is whether dark matter is hot or cold, this means that the computer program only had to be run twice, once for hot dark matter and once for cold dark matter. The results of the two simulations were then compared to the observed large-scale structure of the universe. The results of the comparison were clear. The large-scale structure of the present universe was formed in the presence of cold dark matter. All hot dark matter models are thus ruled out. In particular, neutrinos cannot be dark matter.

We are now out of suspects. So instead of lining up the people we know, we will try to make a composite sketch of our criminal and then see if we can find it out there in the world.

Dark matter is nothing we have encountered before, so it must therefore be some kind of exotic unknown particle. If the idea that most of the matter in the universe is some wholly alien stuff seems disturbing, some cheer can be obtained by buying a helium-filled balloon. It is not so much the balloon itself that is encouraging (despite its perky floating and bright colors), but the difference between the state of our knowledge of helium now and what we knew a little over a hundred years ago. Nowadays helium is a completely commonplace substance, present at almost every child's birthday party. Helium as an element is a basic part of our understanding of chemistry; it is the second lightest chemical element. It is also an important component of the ordinary matter of the universe, making up about one-quarter of the mass of that ordinary matter.

But just a few generations ago, helium was an astrophysical mystery, much as dark matter is today. Its spectral lines had been seen in the Sun, but it had never been detected in any other way. It is often this way in science. Yesterday's mystery has become today's commonplace, and so we have reason to hope that today's mystery will become one of the garden-variety items of tomorrow. In a hundred years, it might be possible to buy dark matter down at the local market—for what use we cannot yet imagine, but it is unlikely anyone imagined helium-filled balloons while staring at the solar spectrum.

Why would you want such an abstruse substance as dark matter? That depends on whether or not it has any uses. Today we buy radios, televisions, cell phones, GPS locators, and so on. None of these things could exist if electricity, magnetism, and light were not all aspects of the same thing. And we would not have any of these devices if James Clerk Maxwell had not figured out and written down four simple but abstruse equations that embodied this interconnection between three hard-to-catch aspects of the universe. It is almost impossible beforehand to determine what discoveries will end up making their way down the floors of science into everyone's apartments.

However, this still doesn't answer the question "What is dark matter?" At this time no one knows the answer to that question. Nonetheless, a plausible guess can be made based on what has been learned so far. This is still just a guess, a good guess, but it could easily be wrong.

Neutrinos have all the properties that dark matter should have except for one: they are hot dark matter rather than cold dark matter. This property of being hot dark matter has to do with the smallness of the neutrino mass (they were small masses exposed to a lot of energy so they started moving really fast). Thus a plausible dark matter candidate would be a particle like the neutrino in the sense that it interacts only weakly, but unlike the neutrino in the sense that it has a large mass and hence would have been slow at the critical time. This is the sketch of our unknown culprit. We could call it John Doe, but that doesn't sound scientific enough. Or someone could make up a name based on its properties. It's a Weakly Interacting Massive Particle, known by the cute acronym WIMP. (Individual scientists may have good senses of humor, but they seem driven to use bad humor in their work, not that they're alone in this.)

Notice that we are now multiplying entities, creating a new hypothetical particle, but we are doing so out of necessity. Scientists hypothesized WIMPs because they ran out of all the entities they knew of and still could not explain dark matter. They made minimal claims about the new particle, assuming it to be neutrino-like but heavy and therefore slow at the proper time. So though they are adding an entity, they are doing so carefully. They have to add it because in this case their parents didn't buy the Christmas presents and Santa Claus is a wimp.

At this point, WIMPs are the most plausible dark matter candidate. What remains to be seen is whether they really exist and if so in enough quantity to be dark matter. To settle this issue there are various experiments that try to detect WIMPs. Since a WIMP is like a neutrino, we might guess that the issues involved in their detection are similar to those used in earlier neutrino detectors, and indeed this is right. As the solar system moves in its orbit around the galaxy, Earth is presumably moving through a sea of WIMPs (assuming this WIMPy theory is true). However, each WIMP interacts so weakly that its probability of colliding with any given atom is very small. As with neutrino detectors, a successful detection apparatus needs bulk, sensitivity, and insulation.

When a WIMP is captured by an atomic nucleus, the nucleus recoils, much like a person stumbling backward after catching a heavy medicine ball. That nucleus then collides with various atoms in the detector. Depending on the material of the detector, the nucleus may knock some electrons out of atoms, a process called ionization. Or if the detector is made of a type of material called a scintillator, then the collisions of the nucleus with atoms can cause tiny flashes of light. In addition, the energy of the collisions between a recoiling nucleus and other atoms will cause a tiny rise in the temperature of the detector. The strategy, depending on the type of detector, is then to detect the tiny numbers of electrons, or the tiny flashes of light, or the tiny rise in temperature.

Such an experiment must be equipped with exquisitely sensitive detectors so that even the small amount of energy deposited by a single WIMP can lead to a detection. To have this sensitivity, each individual detector is fairly small. However, the overall detecting apparatus achieves bulk by consisting of a large number of individual detectors. The detector must be insulated from other things that could be mistaken for a signal. For example, as with neutrino detectors, the detector must be insulated from cosmic rays by burying it deep underground. One experiment called DAMA (for DArk MAtter) is in a tunnel through the Gran Sasso Mountain near Rome. Another called CDMS (for Cryogenic Dark Matter Search) is in an old iron mine in Soudan, Minnesota. Where the experiment measures the tiny rise in temperature due to the WIMP energy, the detectors must be cooled to within a fraction of a degree above absolute zero. At the time of the writing of this book, WIMPs have not yet been detected. More sensitive and bulky detectors are being built. It may be that within a few years WIMPs will be detected and the case of dark matter will be closed.

But we do not yet know this. Until detected or ruled out, the nature of dark matter is one of the subjects under construction in the three-tiered universe. Our criminal remains at large. Here endeth the overlong criminality joke.

CHAPTER 11

All the Way Out

In our dark matter hunt, we have relied on the only thing we can easily measure about dark matter: gravity. We're still stuck with that, but gravity does a lot more than affect the motions of galaxies and clusters of galaxies (as if that weren't enough). Gravity, because it affects the shape of spacetime, affects the expansion of the universe. One might therefore think to measure the amount of matter in the universe by using the law of gravity and the expansion of the universe. In this way the biggest "How much of it is there?" question was asked: How much of everything is there, everywhere? In answering this question, physicists fell into an even greater mystery than dark matter: dark energy.

We would like to do some measurement involving the expansion of the universe that would allow us to calculate the density (mass per unit volume) of the universe. This would give us an idea of how much stuff there is in any part of the universe. Unfortunately, general relativity is not quite this simple; it is a theory not only of gravity but also of the geometry of space and time. In other words, gravity is not simply, as Newton envisioned it, a force affecting objects, but an integral part of the shape of space and time.

Space in Einstein's universe is not an endless sameness where the shortest distance between two points is a straight line. It is a lumpy, bent shape, as we mentioned. It is sometimes analogized as a rubber sheet bent down by each mass on it. As the universe expands, the position and relative closeness of the lumps changes so the bending of the sheet changes. The shape of the universe changes as it grows.

Two equations describe the expanding universe. One relates the square of the Hubble parameter to a combination of the density and the curvature of space. Curvature is a mathematical measure of how much a thing bends at a given place. By "the curvature of space" we mean what the average curvature is, or how much space generally bends rather than how much it is bent in a particular place. The other equation relates the rate of change of the expansion of the universe to a combination of the density and pressure.

When we say that an equation relates one quantity to another, we mean that both quantities are involved in the equation, in such a way that changing one of the quantities means that the other will also change. Here are a couple of examples we've seen:

- $E = mc^2$ relates energy to mass. If mass increases, energy increases. Double the mass, double the energy. This is called a linear relationship.
- $F = GMm/r^2$ relates gravitational force to both masses and to the distance between them. If either mass increases, so does the force; if the distance increases, however, the force decreases. The masses have a linear relationship to the force. Double a mass, double the force. But the distance has what is called an inverse-square relationship. If you double the distance, you divide the force by four.

Back to the equations under discussion. At first sight, the presence of pressure as a source of gravity seems very strange. In our treatment of stars, we have considered pressure as a force opposing gravity. Now, however, we will have to be a bit more careful and sophisticated. The air pressure at sea level is about fourteen pounds per square inch. It is as if a fourteen-pound weight were pressing on every square inch of your body. This sounds like a lot when one considers the large number of square inches on the outside of a human body. Why aren't we crushed by this pressure? The reason is that what we feel as pressure is not absolute pressure, but pressure differences. The pressure inside our bodies is the same as the pressure outside, so we feel nothing (except during those brief times ascending and descending in an airplane when the pressure has not quite equalized). A car's tires can hold it up not because the tires have air pressure but because the air inside the tires has a larger pressure than the air outside the tires.

Similarly, a star holds its shape against gravity not simply because its contents have pressure, but because that pressure gets larger the closer one gets to the center of the star. It is these pressure differences, not the pressure itself, that hold the star up. However, in general relativity it is pressure itself, not pressure difference, that acts as a source of gravity. This is one of the respects in which general relativity is more complicated than Newtonian gravity, where only density of matter acts as a source of gravity. In the weak-gravity and slow-motion conditions of the solar system, the gravitational effects of pressure are much smaller than those of density; but for the expansion of the universe as a whole, it turns out that they are not.

Neither of the two equations for the expansion of the universe is completely satisfactory for our purpose, which is calculating density. If we measure the Hubble parameter and use its square to find the combination of density and space curvature, how do we know what part of that is density and what part is space curvature? To find that out, we need to separate these two quantities.

This process of separation is a tricky one. Mathematically, one separates quantities in an equation by finding some other means of measuring them. For example, if we go back to $F = GMm/r^2$ and we want to know how much of the force on an object is caused by the mass of the object affecting it and how much comes from the distance, we need to be able to independently find one or more of these values. We might, for example, use parallax to find the distance and then we could use that to determine the mass.

We can do this kind of separation in the equation relating density and curvature—or at least quantify our ignorance—by figuring out what the density would be for a given value of the Hubble parameter if space were flat (that is, not curved). This density is called the critical density. Then we can define a quantity denoted by the Greek letter omega (Ω) to be the ratio of the actual density to the critical density. In other words, to separate density and curvature, we figure out what the density would be if there were no curvature and then let omega be the actual density divided by this critical density. Omega then gives us a measure of the curvature. If the space curvature is zero, then the density is equal to the critical density and omega equals 1.

Once the Hubble parameter is known, the question of the amount of matter in the universe then becomes "How large is omega?" Measurements of clusters of galaxies show that they contain regular and dark matter with omega equaling 0.3. One might therefore guess that omega is 0.3 for the universe as a whole and that there is therefore a substantial amount of space curvature. However, there is a cosmological theory called inflation that has as one of its consequences that omega is 1. Another consequence of inflation is a prediction of the nature of the temperature fluctuations in the CMB. This prediction has been verified by careful observation of the CMB. One might then regard this as confirmation of (or at least strong evidence for) inflation and be inclined to also believe that omega equals 1. If omega is 1, then much of the matter is uniformly distributed in the universe instead of being clustered and lumped. If this is the case, there is a substantial amount of matter out there that is not clustered in galaxies.

Recent detailed observations of the galaxies, clusters, and voids that comprise the large-scale structure of the universe, as well as recent extremely accurate observations of the CMB, do lend considerable support to the idea that omega is 1 and therefore that there is a large "unclustered" component of the mass of the universe. This view of things says that not only is there dark matter in the galaxies, but there's some other stuff between the galaxies that makes space relatively flat.

So from a roundabout approach of trying to figure out density, we are led to yet another divergence between what we measured in one situation (omega = 0.3 in galactic clusters) and what we infer from a theory that has independent evidence (omega = 1), and we are pushed to conclude that there is yet again an unknown thing that shows why we get these divergent results.

We might hope to use the other equation mentioned above to try and make sense of this, the equation involving the expansion of the universe. But examining this equation shows us something else that is strange. This equation relates the rate of change of the expansion of the universe to a combination of density and pressure. At first this equation seems to suffer from the same difficulty as the previous one, the difficulty of separating the components. If we measure the rate of change of expansion of the universe, how do we know how much of it is due to density and how much to pressure?

Fortunately, for a gas where the particles are moving much slower than the speed of light—which applies both to ordinary matter and the dark matter in galactic clusters—the contribution of the pressure in this equation is negligible. It was natural for cosmologists to assume that the pressure would be negligible and therefore that if they simply treated the pressure as zero, a measurement of the rate of change of the expansion of the universe would yield a value for the density.

You may ask why density and rate of change of expansion have anything to do with each other. This can be seen if you consider what happens when you throw a baseball straight up. Gravity slows the baseball down, so that one second after it has left your hand, it is moving more slowly than it was half a second after it left your hand. As gravity pulls on it, the ball decelerates. Similarly, we might expect the gravity of every object pulling on every other object to slow down the expansion of the universe so that at later times the expansion is slower than it was at earlier times. That is, we expect there to be a deceleration of the universe; and the higher the density, the higher the deceleration (more stuff slows things down faster).

We measure this deceleration essentially the same way that we measure the Hubble parameter, by measuring the distance and velocity of objects at great distances from us and making a plot of velocity versus distance. Recall that the farther out we look, the longer it has taken the light to get to us and therefore the further back in time we are looking. These earlier objects are seen at a time when the universe was expanding at a different rate. We can plot the distance and velocity of objects, and since longer distance is further back in time, we can see the objects spreading out at different rates as we get further from the present; a few calculations allow us to use this plot to figure out the rate of change of expansion. Cosmologists expected that the shape of such a plot would allow a measurement of the deceleration of the universe, and using the equations discussed before, felt it would be possible to deduce the density of the universe.

This sort of measurement was made in 1998 by two large international teams of astronomers called the High-z Supernova

Search Team and the Supernova Cosmology Project. Before we turn to the results of the measurement, let's look at how it was done (you know the thing we keep pushing, the "How do they know that?" question). The velocities are measured as usual using the Doppler effect. But how are the distances to be measured? So far we have considered in detail only the measurement of the AU and the measurement of the distances to nearby stars by using the method of parallax. We need to measure distances far beyond the ability of this method.

Somewhat larger distances can be measured by using the so-called moving cluster method. This uses a property of perspective drawing (the rendering of a three-dimensional world on a two-dimensional canvas) called the vanishing point. In the three-dimensional world, parallel lines never meet, but an artist drawing a picture of those parallel lines will make a place on the canvas where they meet called the vanishing point. If you look at any drawing of a road going off into the distance, the sides of that road are made to meet in the vanishing point. But of course if you actually walked down the road, the sides would not come together.

Stars often occur in clusters of many stars where the cluster is bound together by the gravity of all its stars. The whole cluster is moving in the same direction, and therefore as seen on the "canvas" of the sky, it is heading for the vanishing point. By taking a picture of the cluster at two different times and drawing lines connecting the earlier position of each star to its later position and then extending those lines, we can find the vanishing point (handy for artists who like to draw starscapes). From these measurements, we know the angular size of the cluster (how large it appears in the sky) and we know how fast it is approaching the vanishing point. But as with all perspective drawings, we don't know the scale. (An object in a perspective drawing is rendered the same size as an object of the same shape that is twice as large and twice as far away.) One more measurement suffices to set the scale, a Doppler measurement of the speed of the cluster. With this and the usual few calculations we don't bother to do here, we will find both the actual size of the cluster and its distance.

There are other methods of measuring distance that have to do with the relation between the distance and two quantities called the apparent luminosity and the absolute luminosity. "Apparent luminosity" means how bright an object looks to us, and "absolute luminosity" is how bright it really is. A 100-watt lightbulb has the same absolute luminosity whether it is one foot away from us or one thousand feet away. However, at one foot it has a much higher apparent luminosity; it's blinding up close to our eyes, but barely visible a long way off.

For our purposes, the important thing is that (a) given any two of the three quantities—distance, apparent luminosity, and absolute luminosity—we can deduce the third; and (b) telescopes are instruments that measure apparent luminosity (because they show the brightness of the object at the distance between it and the telescope). So for any object we can see in a telescope, if we somehow know its distance, we can find its absolute luminosity, and if we know its absolute luminosity, we can find its distance. These two facts allow astronomers to find what are called standard candles and to use them to measure enormous distances.

This works as follows: First astronomers examine the stars whose distances have been found using either parallax or the moving cluster method. From the distance and the telescope measurement of apparent luminosity, the absolute luminosity of these stars is found. Of those stars, the astronomers need to find some that are (a) very bright and (b) have some other property by which stars like them can be recognized and which is related to their absolute luminosity. The first property is needed so that the star can be seen in a telescope at a large distance. The second is needed so that when this same sort of star is seen at a much larger distance, its absolute luminosity will be known and therefore (using the telescope measurement of apparent luminosity) its distance can be deduced.

This last needs a little explanation. Remember that two stars of the same kind (same mass, age, and so on) will have basically the same absolute luminosity since they are both burning the same substances at the same rate. If we can identify one star as being basically the same as another, we can assign it the same absolute luminosity, then combine that with the apparent luminosity and so figure out the distance to it.

The standard candles that Hubble used to find the distance to the galaxies were variable stars. Recall that a variable star is one that does not have a steady brightness, but instead becomes brighter, then dimmer, then brighter, then dimmer in an ever-repeating sequence. Variable stars are very bright, and there is a relation between the period of the variable star (the time between two successive brightenings) and the absolute luminosity of the star. This means that for a variable star seen in another galaxy, the period and apparent luminosity can be measured in a telescope, and from these the distance to the star, and therefore the distance to the galaxy that contains the star, can be deduced.

Once the distance to a galaxy is known, so is the distance of any object in that galaxy. If any object in the galaxy—or even the galaxy itself—is a suitable standard candle brighter than a variable star, then it can be used to measure still farther distances. This method of levering oneself up step-by-step to ever-larger distances using ever-brighter standard candles is called the cosmic distance ladder.

The most important standard candles for our present inquiry are objects called type Ia supernovae. Supernovae is the plural of supernova (who says Latin is a dead language?). These supernovae are different from the supernovae discussed in the previous chapter, which were the precursors to neutron stars and black holes, and whose complete designation is type II supernovae. This arcane notation and the attendant confusion come from the detected universe. Recall that the terminology of the detected universe has to do not with what things are but with how they appear to us. As we have seen in the case of neutron stars and pulsars, the terminology of the detected universe can create confusion by having different names for things that are really the same but that appear different to us. Correspondingly, the terminology of the detected universe can create even more confusion by using the same or a similar name for objects that are very different but that appear the same or similar to us.

A good example of the process of classifying can be found in a common way people learn about the insect world. One often begins with a naive term for a broad class of things like "icky creepy-crawly things." After some study, one might note that there are "icky creepy-crawly things" with legs and those without. One might then divide this category into "bugs" and "worms." Looking at bugs, one might then note that some have six legs (insects) and some eight (spiders/arachnids). One then either retires satisfied or goes into entomology.

From bugs back to the stars. A supernova is a stellar explosion. Its name in part derives from the Latin word for "new" because a star that is ordinarily too dim to be readily seen can become easily visible through the enormous increase in brightness from the explosion. A "new" star is seen where none was seen before. The star's existence is not new, but its appearance is. Observers can distinguish between different types of stellar explosions by looking at their spectra, since the exploding stars have different chemical content. The spectra can be broadly distinguished as type I, in which hydrogen lines are absent, and type II, in which hydrogen lines are present (six legs and eight in the bug example above). Further differences among different type I spectra lead to the classification as type Ia, Ib, and Ic (ants, bees, beetles—we're going to stop with the bug metaphor now). A type II supernova is the explosion of a star caused by the collapse of its iron core. Since the envelope of the star contains plenty of hydrogen, the spectrum of the explosion has hydrogen lines.

A type Ia supernova must then be some sort of star that explodes and that does not contain hydrogen. What is it? And why does it explode? Detections determined that a type Ia supernova is a white dwarf star in a binary system. The white dwarf is accreting matter from its companion, and eventually it accretes enough matter to drive it to the Chandrasekhar limit. A white dwarf is made mostly of carbon and oxygen. When it approaches the Chandrasekhar limit, its gravity makes it collapse. This collapse heats the contents of the white dwarf to the point where they undergo sudden and complete nuclear fusion that blows the star apart. Just as ordinary stars are large, slow fusion reactors, a type Ia supernova is a large fusion bomb, releasing more energy in its explosion than the average star does during its entire lifetime.

A type Ia supernova thus certainly satisfies one of the criteria for a standard candle: It is very bright and can therefore be seen at an extremely large distance. But it also satisfies the other criterion: All type Ia supernovae have nearly the same absolute luminosity. The reason for the sameness is that any type Ia supernova is simply the explosion of a white dwarf whose mass is at the Chandrasekhar limit. To oversimplify a bit, all type Ia supernovae are the same since they are effectively the same kind of stellar explosion starting with the same materials and with the same mass, and therefore they have almost the same brightness. The reason for the qualifying word "almost" is that type Ia supernovae differ slightly, and this makes slight differences in their absolute luminosity. Nonetheless, enough features of the explosion connected with this difference in brightness can be detected and accounted for so that observations of a type Ia supernova can be used to determine its absolute luminosity. Bright and uniform in nature, they are excellent standard candles.

CHAPTER 12

Dark Energy, Antigravity, and Einstein's Fudge

Armed with the means of determining position and velocity of whatever they needed to look at and fully expecting to measure the deceleration of the universe, the two international teams of astronomers in 1998 performed observations of many type Ia supernovae in distant galaxies, finding the speed using the Doppler effect and the distance through the use of the supernovae as standard candles—through which they were able to make the desired plot of distance and velocity and carried out the calculations of universal expansion (which is what they wanted in the first place; sometimes you have to go down a lot of side trips and pathways to solve these things). Much to their astonishment—and everyone else's—instead of deceleration, they found acceleration. The expansion of the universe is speeding up.

At first this doesn't sound so bad. After all, the relevant equation depended on both density and pressure, and the expectation of a deceleration depended on the notion that the pressure was small enough to be neglected. There must be some actual pressure out there. However, it turns out that the gravity due to positive pressure also causes deceleration. The only way for there to be an acceleration is if it is caused by matter that has a very large and *negative* pressure. Furthermore, since gravitational attraction causes deceleration, one can only get acceleration from a type of matter that is gravitationally repulsive. Whatever is causing the acceleration of the universe has a large negative pressure and causes gravitational repulsion.

Now, while antigravity is a staple of science fiction, the truth of the matter is that up until the supernova observations of 1998, there was no reason to believe that such a thing existed. (Incidentally, the first term for antigravity was "levity," though this term has fallen or risen from science into humor.) Newton declared that every object attracts every other object based on the amount of mass it contains. The assumption was that gravity always pulled things toward each other (unlike, say, the electromagnetic force in which like charges repel each other). So whatever this funny stuff is that has the negative pressure and gravitation, it is yet another kind of matter that we have never encountered before.

Physicists do not call this stuff dark matter, since whatever dark matter is, it seems to have negligible pressure and therefore attractive gravity. Instead they call this stuff dark energy. Whatever dark energy is, there is a lot of it. With omega equaling 1 and with an omega of only 0.3 in baryonic matter and dark matter put together, that leaves an omega of 0.7 in dark energy. In other words, whatever dark energy is, there is more of it in the universe than of all the other stuff put together.

In the last few sections, we have seen the importance of our kind of matter (baryonic matter) be displaced first by dark matter (of which there is a lot more) and then by dark energy (of which there is even more than that). This is not the first time our inherently egocentric views have been slapped upside the head by science. The discovery that Earth is not the center of the solar system, then the further discovery that the solar system is just one among many star and planet systems, and then that the Milky Way is again only one of a vast multitude of galaxies can give a bit of a jolt. But it's a good kind of jolt; it reminds us that not everything revolves around us. On the other hand, relativity can be used to give us the opposite feeling. The whole universe can be seen as centered around each and every one of us. Oddly enough, this paragraph has little to do with science as it is, and a great deal to do with science as it is used to tell us about ourselves. But that is not a scientist's use of science; it's a writer's use. Writers, not just science fiction writers, pull upon science as a source of metaphor. It's very useful but rarely accurate to the science called upon. We'll discuss this a little more in the final chapter.

Back to the strange stuff in the universe. Peculiar though dark energy is, its discovery actually solved what threatened to be a crisis in cosmology having to do with the age of the universe. Recall that a first estimate of the age of the universe was found by dividing 1 by the Hubble parameter, giving a value of about 14 billion years. This is crude because it does not take into account that H changes with time. A more sophisticated estimate using the deceleration that would come from a universe where omega equals 1 and containing only ordinary and dark matter leads to an age of only two-thirds of 14 billion years, or about 9 billion years. This is disturbing, because astronomers can estimate the ages of some of the oldest stars, those in so-called globular clusters, and they appear to be a bit older than 9 billion years. It would indeed be embarrassing if astronomers were to claim that the universe is younger than some of its stars.

At the time just before the supernovae measurements, there was enough uncertainty in the estimated value of the Hubble parameter and the globular cluster figures for the age of the oldest stars that one could not say for sure that they were in contradiction; but they certainly seemed on a collision course in the sense that a bit more accuracy in either was likely to lead to a contradiction. Dark energy made this whole problem disappear. Because the universe is accelerating rather than decelerating, the estimate of the age of the universe changes drastically. The current estimate of the age of the universe, about 13.7 billion years, is comfortably longer than the estimated age of any star.

Although helpful, dark energy is extremely strange, so strange that you might think that a small army of physicists is working night and day to think of a type of matter strange enough to be dark energy. And they are. But it might surprise you to know that Einstein thought of such a type of matter back in 1917. Einstein began by tackling a problem whose solution had eluded Newton: the behavior of a universe where matter is uniformly distributed with the same density everywhere. In Newtonian gravity, each piece of matter is acted on by the gravitational forces of all the other matter. To get the total force on any piece of matter, one must add up all the forces due to the other pieces—that is, one must do an infinite sum. An infinite sum is not necessarily an insurmountable problem. A standard result of mathematics is that 1 + 1/2 + 1/4 + 1/8 + 1/16 and so on forever adds up to 2.

Note that many infinite sums, even seemingly small ones, add up to infinity. 1 + 1/2 + 1/3 + 1/4 + 1/5 et cetera is infinite. There are also infinite sums with terms of both positive and negative sign like 1 − 1 + 1 − 1 + 1 − 1 and so on that have no definite answer. Depending on how the terms are grouped, you can make it look like the answer is zero (by using the grouping [1 − 1] + [1 − 1] + [1 − 1] . . .) or 1 (by using the grouping 1 + [−1 + 1] + [−1 + 1] + [−1 + 1] . . .) or any other answer that you care to choose. For an infinite sum to make sense, the numbers to be added must get smaller sufficiently fast. The mathematical term for this is convergence.

It turns out that the gravitational forces due to the matter in an infinite universe do not get small fast enough for the infinite sum to make sense. Newton's sum has positive and negative terms. There is a grouping of the terms—the one Newton chose—that makes the answer appear to be zero; but there are also other groupings that make the answer appear different. Now, with the modern knowledge of infinite sums, we know that there simply is no answer to the question that Newton posed about the behavior of a universe in which matter is uniformly distributed and uniformly dense. However, Newton did his calculation in a time before these properties of infinite sums were worked out and mistakenly thought that all the forces would cancel each other out and that the matter would not move.

Einstein noted that his own theory of general relativity, unlike Newtonian gravity, allowed the possibility of a curved space and in particular allowed a space that was curved enough that the total volume of space was finite. The particular space that Einstein considered is called a three sphere. To understand a three sphere, imagine that one lived on a regular sphere (called a two sphere in math). That is, imagine living on the surface of a ball so that whatever direction one went—north, south, east, or west—one would eventually return to the same spot. Weird way to live, huh? To get a three sphere, imagine that up and down also behaved like north, south, east, and west. Go up all the way or down all the way, and you end up back where you started. In other words, a three sphere is a space where whichever way you head out, you eventually return to where you came from. Such a universe could only contain a finite amount of matter, since it only has a finite space to contain it, so Newton's problem of the infinite series would not arise.

Like Newton, Einstein thought that in his three-sphere universe the answer to the uniform density-gravity problem would be that there is no net force on any of the matter and that therefore none of the matter in the universe would move. However, that answer was not consistent with Einstein's own field equations of general relativity. And so to get the answer that he wanted, Einstein changed his equation by adding a new term to it, a cosmological constant. The "antigravity" of the cosmological constant cancels out the gravity of the matter and allows the matter of Einstein's three-sphere universe to be unmoving.

This did not solve the problem Einstein was considering, because this balance between gravity and antigravity is very delicate. For there to be no force, the matter density would have to have the exact value needed to balance the cosmological constant. If the matter density were ever-so-slightly smaller, then the universe would expand forever; and if the matter density were ever-so-slightly larger, then the universe would collapse in a big crunch.

Even more importantly, the universe *is* expanding, as found by Hubble several years after Einstein proposed the cosmological constant. When Hubble's discovery of the expansion of the universe was announced, Einstein realized that his theory of the cosmological constant was wrong, and he went back to his original version of general relativity without a cosmological constant. He later called the cosmological constant his biggest blunder.

Here the blunder consisted not merely in being wrong, but in a gigantic missed opportunity. The expansion of the universe is a consequence of general relativity without the cosmological constant. Einstein had the opportunity to predict this expansion years before it was discovered by astronomers. Instead, he chose to change his theory in an attempt to get rid of its startling prediction. While some people take their theories too seriously, Einstein apparently did not take his theory seriously enough. It is also possible that the idea of the expanding universe did not seem comfortable to his mind and that he went the way of comfort. Even the greatest of geniuses can make mistakes based on how they feel things should be.

Ironically, though Einstein introduced the cosmological constant in an ad hoc way that got in the way of understanding his own work, it was later found that just such a cosmological constant is a natural consequence of particle physics theories of the vacuum (which, if you recall from earlier discussions, is a very strange place indeed).

As we discussed earlier, quantum mechanics says that each particle can exist at various energy states, and there is a lowest such state called the ground state. An atom as an aggregate of particles also has a state of lowest energy, the atom's ground state. A natural question to ask for atoms is "What is the energy of the ground state?" For hydrogen, it is equal to minus the energy an electron gets from a voltage of 13.6 volts (that is, minus 13.6 electron-volts; electron-volt is a unit of energy small enough for the atomic scale). For the vacuum, "spread out" as it is over all space, we want to establish what the ground-state energy density (the energy per unit volume) is. Remember that paradoxically the vacuum is not really empty, but instead consists of pairs of virtual particles popping into and out of existence. Therefore the question "What is the energy density of the vacuum?" may not have the obvious answer of zero.

Recall that according to relativity, fundamental physics is independent of the velocity of the observer; the universe works the same regardless of how the measurer of the universe is moving. If the vacuum ground state also has this property, then it must appear the same to all observers regardless of their velocity. If the vacuum has energy, then for that energy to appear the same to all observers, it must also be accompanied by a negative pressure. This is exactly the kind of uniform negative pressure we're looking for. Using the mathematics of special relativity, it can be determined that the vacuum must behave exactly like Einstein's cosmological constant. Unfortunately, these theories do not tell us how large that cosmological constant would be, so we can't determine from this if the vacuum is itself dark energy. However, it was this realization of the cosmological constant as an energy density of the vacuum that gave rise to the name "dark energy."

The antigravity properties of the cosmological constant are exactly what is needed to produce the observed acceleration of the universe, assuming the values are correct. Though particle physics doesn't predict the size of the cosmological constant, we can use dimensional analysis to guess the size. The ingredients would be Newton's gravitational constant G, the speed of light c, and Planck's constant h, which gives the relation between the frequency of a light wave and the energy of the corresponding photon. (Planck's constant is a vital quantity in quantum mechanics.) There is exactly one combination of G, c, and h that has the units of energy density, so we might guess that the cosmological constant has this value.

Unfortunately, it turns out that this value is about 10^{120} times larger than the observed density of dark energy. This is probably the most spectacular failure of dimensional analysis in the history of physics. In the previous section of the book, we used dimensional analysis to explain the formula for the Schwarzschild radius of a black hole. Dimensional analysis is a very general and extremely powerful tool in physics. Though it only gives estimates for physical quantities, in general those estimates are not strikingly different from the actual values *unless* some crucial piece of the relevant physics is not properly understood and therefore not properly applied to the dimensional analysis calculation. The spectacular failure of dimensional analysis in the case of dark energy indicates that there is some important part of dark energy physics that we don't yet understand.

Before the observation of dark energy, most particle physicists thought that the cosmological constant was zero and that it was a task for particle physics to find the reason why it was zero. Sidney Coleman, a theoretical physicist with a flair for both elegant mathematics and elegant writing, even wrote a paper in 1988 with the title "Why There Is Nothing Rather than Something: A Theory of the Cosmological Constant."

Now the task appears to be either to explain why the detected cosmological constant has this outrageously small value (from the point of view of dimensional analysis, it is outrageously small) or to find some theory of dark energy other than the cosmological constant. Several such theories have been proposed, but it is not clear whether any one of them is more plausible than the others, or whether any is more plausible than a simple cosmological constant.

Note that while the investigation of dark matter began with several usual suspects (all now ruled out), dark energy began with only one: cosmological constant/vacuum energy. In science, even more than in police work, it is dangerous to begin an investigation with only one suspect. Therefore, theoretical attempts to "round up the unusual suspects" are appropriate and are being done.

The present universe—consisting of baryonic matter, dark matter, and dark energy—presents us with some mysterious coincidences. The first coincidence has to do with a comparison of ordinary matter and dark matter. Recall that the baryon-to-photon ratio is about half a billionth. This means that there is only one baryon for every 2 billion photons. Thus any theory of baryogenesis (the production of baryons in the early universe) must account for this small number. Similarly, there must be another very small number that is the dark matter particle-to-photon ratio, and any theory of dark matter must account for this small number also.

At the present time, the dark matter is about seven times as dense as baryonic matter. However, since both dark matter and baryonic matter densities change in the same way with the expansion of the universe, this ratio of dark matter density to baryonic matter density was also about seven throughout most of the history of the universe. Thus two extremely small numbers (baryon-to-photon and dark matter-to-photon ratios) characterizing two very different processes somehow yield densities that are not that different from each other. Whatever dark matter is, why is there approximately seven times as much of it as there is of baryonic matter when there are so many more photons than either of them? This may just be a coincidence for which there is no explanation. But it seems plausible that any theory of the formation of baryonic matter and dark matter would explain this coincidence, indicating some hidden harmony between dark and baryonic matter.

A different type of coincidence comes from the comparison of the densities of dark matter and dark energy. At the present time, dark energy density is a little more than twice dark matter density. Now suppose that dark energy is a cosmological constant. Then the density in dark energy does not change. However, the expansion of the universe dilutes dark matter particles and makes their density go down. Eventually, the universe will expand enough that the density in dark matter (and in baryonic matter) will be a tiny fraction of the density in dark energy. Correspondingly, at early times, the density in dark energy was a tiny fraction of the density in dark matter. Thus it appears that we live at a very special time in the history of the universe, the epoch when neither dark matter nor dark energy is negligible. Here, too, this may just be a coincidence without an explanation; but we might certainly hope that a theory of dark energy explaining this coincidence could be found.

The above paragraphs hint at the kind of thinking that goes on in the depths of theoretical physics. When theories are being created, the theoreticians have to try to discern what it is the theory will have to account for. A theory of particle physics has to account for atoms, for example, but need not concern itself with the behavior of llamas toward sheep. In trying to find a theory for something as hard to get a handle on as dark energy, scientists have to find what aspects of the universe seem to need explaining in this area and try to create a theoretical structure that will explain it.

But though cosmologists continue to create and explore theories of dark matter and dark energy, in the end these issues are likely to be made more clear not in the theoretical but in the detected universe. Direct dark matter searches are ongoing. If they detect dark matter particles, this will give us needed information about the properties of dark matter. Similarly, more detailed studies of the history of the expansion of the universe will yield more information about the acceleration of the universe. This can be used to find properties of dark energy and in particular to see whether it is a cosmological constant.

The subject of this chapter is definitely unknown physics, but only partly so. Though some of the theories of dark matter and dark energy are somewhat speculative, the fact that these substances exist is not. This makes dark matter and dark energy somewhat different from other exotic areas in the frontiers of physics. If Mark Twain were alive today, he might well dismiss such things as superstrings and extra dimensions as the products of the overactive imaginations of theorists. But dark matter and dark energy cannot be wished away in this manner. Their existence is solidly established by the methods of the detected and theoretical universes, as is the fact that they cannot be ordinary baryonic matter. The search for the nature of dark matter and dark energy is one of the most exciting questions of our time. The answer, if it is found before the turn of the next century, is sure to be regarded as one of the triumphs of twenty-first-century physics.

We have reached the working edge of our three-tiered universe, the place where the construction is ongoing. As warned, this chapter has less of knowledge and more of the noise that comes from the creation of knowledge. Some might complain that a book on science should be about what is known, not about the process of finding it out. But as we move into the next chapter it will become clear that even if we live in the lower established stories of the buildings of science, the noise of creation makes its way down. Those who live on the lower stories need to know what all that noise is about, since sometimes our lives are the nails to the hammering.

We will start where we now stand: at the edge of the universe from whence we shall fall back to Earth, where we began.

A Step Back to Earth

Previous page: Earth from space. NASA, 2007.

CHAPTER 13

Road Maps

We don't intend to leave off at the edge of the universe, tempting though it is. That would be like driving someone all the way across the country and then abandoning them, saying, "Find your own way home." At the beginning of this book, we promised a round-trip and we intend to keep our promise, because science is not just an abstract endeavor that keeps scientists off the streets (there's nothing more dangerous than roving bands of cosmologists, unless it's geochemists—those people are scary). Remember that we started out talking about cell phones and X-ray machines. None of these could have been invented without considerable knowledge of science, knowledge put to use in ways that enter people's lives all the time. The point of taking the return trip is to travel down paths of scientific understanding to reach the lives we live. It isn't enough to go out from the perceived universe; we also have to come back in order to make the endeavor of science worthwhile to everyone (not just the aforementioned science gangs).

Before we return, let us take a look at the route we followed to get here. In the introduction, we set out a particular method for looking at science: emphasizing that much of nature is not directly accessible to our senses, but that it can nonetheless be understood through the use of detection apparatus and theory. We also emphasized that understanding of science begins with the question "How do they know that?" The next three sections applied this method to astrophysics and in particular to the understanding of the Sun, black holes, dark matter, and dark energy. In this section, we will return to our broad general themes and consider what our method has to say about sciences other than astrophysics.

What we are about to do is analogous to making a large-scale road map that passes through a country. It would be nice to make a full large-scale map of science, but that would require a book unto itself. This chapter is more like the kind of map one uses to get from one place in one part of a country to another place in another part. Such a map would have marked on it the major highways, the names of regions and cities, and labels for really big geographic features (oceans, large lakes, mountain ranges, and so on). It would also show the route of one's passage, marking the bends and twists, showing the entrances and exits. Such a map is not useful for getting to each particular address in each particular city—for that, one uses small-scale street maps—but for finding one's way across a huge swath of territory, for knowing in general where one is along a route, a large-scale map is invaluable.

The treatment of the last three sections is essentially a detailed road map of a province in the larger realm of science. In this section, we will sketch a map that leads from the science we have been doing, astronomy, to the science that has the largest impact on our daily lives: medicine.

We have three reasons for making such a map: The first is to place the knowledge of the last three sections in context. The second reason concerns something we have emphasized throughout this book—the indirectness of knowledge in science. But since our examples have been from astronomy, it is tempting to think that this indirectness is simply a result of the objects of study being so far away. Surely (one might think) when the things we study can be put on the lab bench and held in our hands, knowledge is not so indirect. Instead, it turns out that the indirectness is simply of a different kind. We will use this chapter to demonstrate that point.

The third reason is to bring to the fore something we have employed and noted in previous chapters: the interconnection of the branches of science. We saw this in the use of a chemical test in discerning the nature of the Sun, in the vital role that quantum mechanics plays in exploring the matter in white dwarfs and neutron stars, and in the fundamental role mathematics plays in all scientific inquiry (actually the relationship between math and science is even more complicated than this, but that's another book).

It is important to realize that the distinctions we make between the sciences are distinctions of human circumstance and convenience. Each of us has only so much time alive, and only so much we can do and learn. We cannot hope in a single lifetime to learn all things and to do all things. Furthermore, we are not all interested in the same things or even with the same aspects of the same things; one child fascinated by the flight of birds may become a pilot, another an aerodynamic engineer, another a paleontologist looking at the connections between birds and dinosaurs. We have neither time nor space to do it all, so we classify things, saying this aspect of the bird in flight belongs to the engineers, this to the biologists, this to the paleontologists, and so on. But the bird does not care—and more to the point, the universe does not care—about the distinctions we make. The same understanding of flight can be reached from any of these directions, and the work of the engineer can help the biologist, and both, in turn, can inform the paleontologist, but only if they all understand that they are working on the same things from different directions.

So while we divide science up for our convenience, we also have to unite it for our convenience, otherwise the biologist will waste a lifetime re-creating the work of the aerodynamicist, and the paleontologist will not notice that the bones left over from a chicken dinner are like the bones unearthed in China from feathered dinosaurs.

To make the map of this kind of interconnection, we will, like a person who starts in their home area and makes the map outward from what they know, begin with the territory we have covered and move to other familiar territory, mapping as we go.

FROM THE STARS TO EARTH

In the discussion of stellar evolution, we touched only briefly on those little hard lumps of stuff orbiting the stars: the planets. In many respects, planets are a side effect of stellar formation, and their significance in the life cycle of the stars is minimal. But from our perspective as the inhabitants of a planet, they matter a great deal.

Theories of planetary formation suffer from a lack of data points. While we have a good selection of planets in our solar system, they were all formed as children of a single star. In terms of formulating a theory of planetary formation, we are in the same difficulty as a biologist who has only one small family of creatures to study in order to determine the characteristics of the entire species. Recently planets outside our solar system have been discovered, but they have all been found by the indirect means touched on in previous chapters. Very little information about these extrasolar planets is known. For the most part, only their masses and the distances from the stars they orbit have been found.

There are theories of planetary creation that fit the available facts, but it is likely that they will be refined or changed as our ability to detect the conditions of other planets improves (the detection in other solar systems of large planets in small orbits has already led to such modifications). The most commonly used theory at the moment is the planetesimal theory. In this theory, the formation of the Sun from the collapse of a cloud of gas was accompanied by the formation of a disk of material in orbit around the young Sun. Like the Sun, that disk was composed mostly of hydrogen and helium (collectively referred to as gas); but there were also small grains of various minerals (collectively referred to as rock) as well as small grains of solid water, ammonia, and methane (collectively referred to as ice). As time went on, the grains of ice and rock collided and stuck, eventually forming larger and larger objects called planetesimals. Eventually collisions between the planetesimals formed the precursors to today's planets. Since the inner part of the solar system was hotter, there was less ice and the resulting planets were more rocky and smaller, too small to gravitationally capture the gas, even though the gas is the most abundant of the ingredients in the solar disk.

In the outer solar system, the planets became large enough to capture gas and so grow to an enormous size. After a time this process of collision and collection produced small rocky planets in the inner solar system, large gas giants in the outer solar system, and many rocky and/or icy planetesimals: the moons, dwarf planets, asteroids, and comets. An early and more violent version of the solar wind then swept the solar system of the remaining uncaptured gas, cleaning the plate, as it were.

The planets of our solar system are here to be studied and, at least in one case, lived on (and so studied in great detail). One of the things that can quickly be observed is that the contents and composition of planets are more complex than that of stars. The whole panoply of chemical reactions is missing from the stars because they are too hot for atoms to come together and form molecules.

But here in the cooler parts of the solar system, chemistry becomes possible. The hundred-odd elements can combine into an arbitrary number of compounds. This is a condition analogous to one we saw before in the early universe. In that time, atoms were not possible because matter was too energetic for nuclei to hold together. When things cooled, relatively complex nuclear structures became possible. This is a repeated and important phenomenon. Under some conditions, structural complexity is impossible, but if those conditions are changed, things can combine in wholly unexpected ways.

If you observed the almost-fluid state of the early universe, you would not likely imagine the possibility of atoms. If you saw that ocean of protons, neutrons, electrons, and all the other subatomic particles bouncing around in the radiation soup, so uninterested in each other except to combine and split apart into other elementary particles, would you guess that if things were cooler they could come together into the beauty and quantum symmetry of the atom?

If you saw nuclei whirling around, fusing together in stars, would you guess at the complexity of chemistry? When carbon was first formed in stars, flying about amid the hydrogen and helium, would anybody have guessed that in that single element lay the potential for graphite, diamonds, plants, animals, and us?

Chemistry as a science exists because there are planets. In these cool places, elements can come together. Chemistry and atomic physics overlap as fields of study because the atom lies on the boundary between them, because the quantum structure of electron shells determines what molecules can and cannot form out of atoms. Atomic physics looks at the components of the structure, studying the atom while acknowledging the potential for molecules. Chemistry concerns itself largely with the next level of structure, when atoms become less themselves and more a part of larger building blocks, though they do not lose their nuclear identities.

It is an irony that points toward a vital truth that because it involves this more complex and larger structure, chemistry is a far older field of knowledge than the theory of atoms and molecules. Chemistry is not the most ancient of sciences (that honor belongs to biology, as we will see), but it is old because chemical reactions occur in a way that can be created and perceived on the human scale. It is not the primal character of knowledge that makes it come early in the history of science, but its closeness to the way we do things. Science stretches out from the perceived universe, and the fundamental determinant of the perceived universe is the limit of our perceptions. Experimentation stretches out from the detected universe, and the limit of experiment is the limit of our hands and our tools.

The experiments that chemistry needed to create a vast plethora of particular reactions required nothing but macroscopic apparatus, usually glass and fire. Chemists both theoretical and practical have been filling up the knowledge of the field since before there was writing, long before they thought of themselves as chemists or alchemists or cooks. Chemistry developed initially as a practical science, a thing created by artisans as well as scholars. Proto-chemists wanted to know how to make good gunpowder, how to create glazes for pottery, how to ferment better beer and wine, and how to dye clothing richer colors.

The previous paragraph seems to place chemistry squarely in the perceived universe, as a science of mixing things together and seeing what happens. This is certainly a part of chemistry; but it is important to remember that chemistry came to concern itself with atoms, and atoms are extremely small. How small? Well, depending on what type of atom it is, the diameter of an atom is from about one- to a few ten-billionths of a meter. Or to put it another way, there are more atoms in a glass of water than there are glasses of water in an ocean. Chemists deal with these tiny distances in much the same way that astronomers deal with huge ones: in scientific notation. One ten-billionth of a meter is written as 10^{-10} meters, which is then given the unit name of 1 angstrom. So we can go around saying that atoms are a few angstroms long, just as we said that Earth is 1 AU from the Sun, but this notation doesn't change the fact that these length scales are vastly different from those of our daily life.

It is sometimes said that the large distances in astronomy make us feel tiny and insignificant. If so, then thinking about chemistry should provide the perfect antidote: compared to atoms, we are huge, gargantuan oafs, gigantic clumsy assemblies of over a billion billion billion atoms.

How then does one gain knowledge of this world of tiny atoms much too small to see? Sophisticated present-day equipment (electron microscopes and atomic force microscopes) have the resolution to see individual atoms. However, much of the detection apparatus of chemistry is of the "mix it together and see what happens" variety. A nice example of this pleasantly mad-science style experimentation concerns acids and bases. Water is H_2O (two hydrogen atoms and one oxygen atom), but in a sample of water a small percentage of the molecules dissociate into H^+ (hydrogen missing its electron) and OH^- (oxygen bonded to hydrogen with an extra electron). This notation may look odd, since + means lacking an electron, and – means having an extra one. The notation represents the total electric charge. Atoms normally have an equal number of protons (charge +1 where the unit in this case is the charge of a proton) and electrons (charge −1) producing a total charge of zero, electrically neutral. But an extra electron means a total charge of −1 and a lack of one electron means a total charge of +1, which are simply abbreviated as – and +, respectively. If you want to know why protons are said to have a positive charge and electrons negative, blame Benjamin Franklin (we're not kidding; it's his fault).

Since each H_2O that splits makes one H^+ and one OH^-, it follows that water contains equal amounts of H^+ and OH^-. Add some appropriate chemical to the water, and it may disturb that balance, leading to an excess of H^+ (in which case the chemical is called an acid) or an excess of OH^- (in which case the chemical is a base). There is a simple test to tell if a liquid is an acid or a base. All that is needed is litmus paper, which contains a chemical that turns pink when exposed to an acid and blue when exposed to a base. Litmus paper is one of the simplest examples of detection apparatus: we can't tell by looking at a chemical whether it is an acid or a base; but dip a piece of litmus paper in it and see what color the paper turns and immediately we know. What if we have an acid and want to know how strong an acid it is (that is, how strong an imbalance between H^+ and OH^- there is)? Since acids and bases have opposite imbalances, mixing them together can neutralize the imbalance in each. To measure the strength of an acid, all that is needed is to take a base of known concentration and then measure how much of the base needs to be added to the acid in order to neutralize it. In this way, the properties of chemicals are measured without ever having to directly deal with the tiny atomic length scale. In fact, detection and neutralization of acids and bases goes back centuries further than the present atomic theory.

The theoretical universe of chemistry is largely concerned with how atoms come together to form molecules and therefore what molecules can be formed and under what conditions. The latter is vital because chemistry is a practical science. Its answers lead directly to new inventions and new determinations of what is safe and what is dangerous. Chemical theory needs rules for how atoms come together to form molecules. We have already encountered the source of these rules: quantum mechanics. Just as quantum mechanics determines the configuration that electrons take in a single atom, so it also determines the configuration that electrons take in two or more atoms.

In molecules, not all the electrons belong to one atom or the other: some electrons are shared by more than one atom. This sharing of electrons, which is called a chemical bond, keeps the atoms together, since neither will stray too far from their shared electrons. To get an idea of what molecules can form, we need to know how many electrons each type of atom tends to share.

An atom consists of protons, neutrons, and electrons; the number of electrons can vary if some are ionized off or shared. So atoms are classified by their nucleons (protons and neutrons). We mentioned previously that two atoms with the same number of protons are considered the same kind of atom (the word "element" means the same kind of atom). If they vary in their number of neutrons, they are called isotopes of that element. This classification by protons is not subatomic bigotry. It is because the number of protons determines the chemical behavior of the atom.

There are over one hundred different types of atoms. The periodic table of the elements lists atoms by how many protons they contain, but then organizes them into categories (the columns of the table) based on how they share electrons. For illustrative purposes, we will concentrate on the four kinds of atoms most vital to life: hydrogen, carbon, nitrogen, and oxygen. These are some of the lightest elements. Using quantum mechanics, it can be shown that hydrogen will share one electron, oxygen will share two, nitrogen three, and carbon four. Helium, while extremely abundant in the universe, does not share electrons and so does not form molecules. The column that helium belongs to (containing helium, neon, argon, krypton, xenon, and radon) is called the noble or inert gases. "Noble" because they're too snooty to associate with other more common atoms, "inert" because they don't do anything chemically. Any political conclusions you choose to draw are purely up to you.

To see what molecules can form, it is helpful to draw diagrams where each atom is denoted by a letter (H for hydrogen, C for carbon, N for nitrogen, and O for oxygen) and each bond is denoted by a line. Such diagrams are shown in figure 12.

To begin with, we can consider molecules made by combining each of the above atoms with hydrogen. Two hydrogen atoms can share electrons with each other to form a hydrogen molecule. In the compact notation of chemistry, this is denoted H_2, where H is the symbol for hydrogen and 2 means that the molecule contains two such atoms.

An oxygen atom can combine with two hydrogen atoms, since the oxygen atom shares two electrons and each hydrogen atom shares one. This molecule is denoted H_2O, the well-known formula for water. Similarly, nitrogen combines with three hydrogen atoms to form the molecule NH_3 known as ammonia, while carbon combines with four hydrogen atoms to form CH_4, which is called methane. It is helpful to think of atoms as individual building blocks and molecules as the things that can be built out of the building blocks. As with any children's blocks, quite large and complicated things can be built, even when there are only a few different kinds of blocks.

For molecules, this can be illustrated using just carbon and hydrogen. Consider a line of carbon atoms, each having one bond with the one next to it. Each carbon in the middle of the chain has two chemical bonds, one with the carbon in front of it, the other with the one behind it, while the carbons at the ends each have only one bond. To bring each carbon to four chemical bonds, simply add the appropriate number of hydrogen atoms, of which there are usually plenty available: two for each carbon in the middle of the chain and three for each of the carbons on the end. This sort of molecule is shown in figure 12.

The simplest such molecule is methane (CH_4), while the next simplest, with the chemical formula C_2H_6, is called ethane. After that there is propane (C_3H_8), then butane (C_4H_{10}). The names of these gases may sound familiar, as will a common term for them,

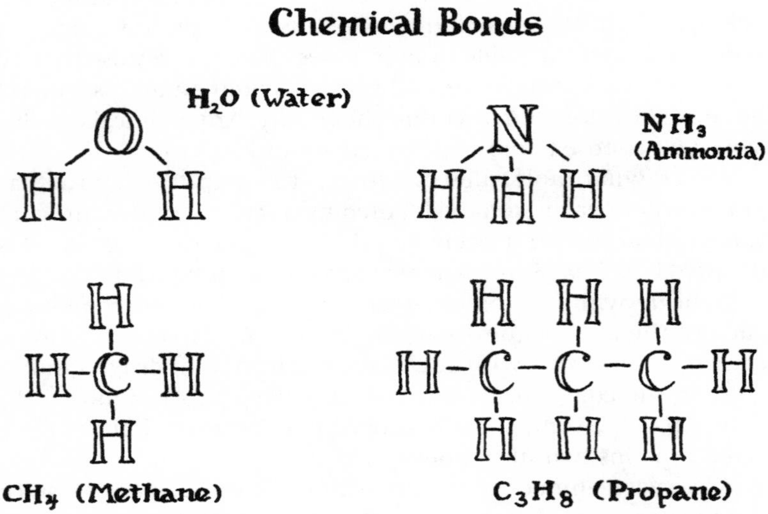

Chemical Bonds

H_2O (Water)

NH_3 (Ammonia)

CH_4 (Methane)

C_3H_8 (Propane)

Figure 12

hydrocarbons. They all burn very well and hence all are used in one form or another as fuel, but they have variations that serve other functions as well.

Since hydrogen has a single chemical bond, in any molecule it can be replaced by anything else that has a single chemical bond. This is one of the critical points of chemistry. Atoms or molecules that have the same bonding properties can be substituted for each other in order to create more complex or simpler molecules, just as building blocks that fit in the same holes can be replaced with each other.

In particular, we usually think of water as an O connected to two H's. But we can just as easily think of it as an OH connected to an H. The OH has a single bond with the H, and therefore OH can have a single bond with anything else. For any molecule we have considered so far, we can replace an H with an OH to get a new molecule. Do this to ethane and you get ethyl alcohol (usually just called alcohol), the "active ingredient" of beer, wine, and liquor. Do it with methane, and you get methyl alcohol; with propane, you get propyl alcohol; and so on. (Be warned, every other kind of alcohol is even more dangerous to your system than ethyl alcohol. Just because something is labeled alcohol doesn't mean you can drink it.)

In one very important respect, however, atoms differ from building blocks. We can hand-manipulate blocks to assemble them block by block in any way that we like. However, atoms are much too small for this sort of manipulation. How then are molecules assembled? If we look back at earlier examples of assembly (such as fusion), we might guess that force and energy were responsible, and we'd be right.

Chemical reactions are powered by brute force. Two or more molecules collide with each other, break apart, and recombine in different ways. An example is the burning of hydrogen to form water. Here two hydrogen molecules combine with an oxygen molecule to form two water molecules. The chemical formula for this process is $2H_2 + O_2 = 2H_2O$. More generally, in a fire various molecules made of carbon, hydrogen, and sometimes oxygen are combined with oxygen in such a way that the hydrogen atoms combine with oxygen to give water and the carbon atoms combine with oxygen to form carbon dioxide (CO_2). This is also what happens when gasoline is burned in a car engine.

Since chemical reactions depend on collisions between molecules, and since the speed of the molecules depends on the temperature, chemical reactions are often sensitive to the temperature at which they take place. This is why a fire must be started usually by another fire (such as a match) or by electricity (as happens in electric sparkers) or by friction (such as a match being pulled across the striking stripe on a matchbox); the temperature has to be raised high enough for the chemical reaction to take place for burning to occur. But it is also why a fire, once started, continues to burn so long as it has new fuel; the energy produced by the chemical reactions keeps the temperature hot enough for the next batch of reactions to continue to take place and so on, as long as the energy and fuel last. This means that much of chemistry is the voluminous lore of what chemical reactions will take place and under what circumstances.

Chemistry, unlike astronomy, is a science that initially concerned itself with the human scale of existence. But, delightfully, this parochial origin did not prevent chemistry from stretching out its hands and helping us to see the universe. Remember, it was from chemistry that astronomers learned to discern the makeup of the stars. This justifies, as we said it would, the large-scale road map of scientific advances. Just as one city or area may provide a product of use to another city or area, so the regions of science trade knowledge and methods with each other. In this trade, all sides gain.

LIFE

If chemistry seemed at first sight a science of the perceived universe, this is even more true of biology. We are surrounded by many easily perceived living creatures: each other, our pets and houseplants, birds, trees, and grass, just to name a few of the most obvious. Knowledge of life has always been extremely important to human beings since we need to consume other living creatures in order to survive, and some of them like to consume us. Furthermore, some of the things we might consider consuming will kill us if we eat them. The knowledge of what plants and animals are good to eat and how to gather or capture them long predates written records, making biology easily the oldest and most practical of sciences. Such hobbies as gardening and bird-watching are present-day amateur biology accessible to all and using the techniques of the perceived universe (though the senses of the bird-watchers are often augmented by binoculars).

However, as with much of science, this sense of comfortable familiarity disappears once one looks more closely. Under the microscope, an enormous surprise is revealed. All life is made up of cells, where each cell is so small that it is invisible to the naked eye. The average animal cell has a length of about ten-millionths of a meter. Or to put it another way, the average adult human body contains about 100 trillion cells. Some living creatures are single cells. Others like us are vast arrays of cells. The cells are grouped together in tissues and organs, with each cell, tissue, and organ specialized for different tasks.

Under an electron microscope, one is further surprised to discover that life and nonlife are not as separate as one might like to think. Living things are largely made up of water and organic compounds. An organic compound is a molecule composed of chains of carbon atoms with other atoms hanging off it, such as those we discussed before. Many of these compounds are immensely complex in structure, but nevertheless they are nothing more than molecules formed by plain old chemical processes.

Life as we understand it arises in the interaction of these molecules. One can think of living things as being combinations of many molecules undergoing an enormously complicated series of self-sustaining chemical reactions. This is called a reductionist viewpoint. One could equally well say that the perceptions of daily life reveal to us a vast array of animals and plants of various shapes and sizes; one would like to examine these living things more closely using various methods of detection.

A close examination of individual cells shows that each cell is undergoing a vast array of complicated chemical reactions, and it is these reactions that allow the cell to do the things it does. In addition, it is found that from time to time a cell divides, becoming two cells, each of which functions in much the same way as the original cell did.

A natural question to ask is "Why are cells so small?" The answer can be found using our old friend dimensional analysis. Since a cell sustains itself through chemical reactions, it must take in chemicals from outside itself to undergo the reactions. It must also expel those products of the chemical reactions that it doesn't need. In other words, like all life an individual cell must eat and it must get rid of waste. Now consider two cells of the same shape but with one twice as large in all dimensions as the other. The larger cell has twice the diameter, four times the surface area, and eight times the volume of the smaller cell. Four times the surface area means that the larger cell can take in chemicals at a rate four times as large as that of the smaller cell. But eight times the volume means that to sustain the same chemical reactions the larger cell would need to take in chemicals at a rate eight times as large as that of the smaller cell. The bottom line is that the task of the larger cell is twice as hard as that of the smaller cell. For a cell, the smaller it is, the more efficient a chemical factory it is.

This line of thought then turns the question around to "Why aren't cells even smaller than they are?" Here the answer is that cells are made of molecules, and a large number of molecules are needed for the cell to do the complicated self-sustaining things characteristic of life. The optimum size for a cell is the smallest it can be subject to the constraint of how complicated it needs to be. Later our discussion of evolution will make clear why cells are at approximately an optimum size.

Many of the structural pieces of the cell, the chemicals of life, are made of molecules called proteins, which are long chains of many small molecules called amino acids connected end to end. Some proteins regulate the chemical reactions of the cell by being catalysts. Here "catalyst" is a term in chemistry for a chemical that participates in a reaction but is neither created nor destroyed in that reaction. The amount of the catalyst affects the rate at which the reaction happens.

In biology, a protein that is a catalyst is called an enzyme. To function, a cell must be able to make its structural proteins and its enzymes. If it could not manufacture the proteins it needed, it would not exist. Without the enzymes, it could not regulate itself, and so would either fail to create the needed reactions or have too many of them and therefore use up its own substance in trying to survive.

How does the cell know how to make its proteins? Before we answer this question, let us look at the question itself, because the form of it contains a dangerous word, "know." "How does it know?" Knowledge is something that arises from our direct experience. To us as thinking creatures, knowledge and volition (acts of will) are inherent in the way we work, so we tend to project the ideas of knowledge and will outward onto the world. Thus we curse the weather and our computers for "choosing" to mess up our lives. Equally we look at something like a cell doing something like building its complex proteins and we think that what it has is "knowledge" of how to do it. It doesn't; what it has is sophisticated mechanisms that create using information stored.

Information, as we saw in quantum mechanics, is not just what we know, but also inherent characteristics of things. We look at the periodic table and see information about the ways the elements behave chemically, but the elements themselves do not look at the periodic table to see how they should act; they just act. What we are about to talk about, the genetic code, is an example of an incredibly sophisticated use of things acting as they act. We have a sapience-centric view that sophistication equals intelligence, but sometimes things just happen in complex ways, and, more importantly, sometimes there are mechanisms that can push aside one means of doing things in favor of another, not because they are acting from volition but simply because one thing does the job in a way that makes it take the place of another thing that does the job less well.

The information needed to make proteins is coded in another molecule in the cell, called DNA. DNA is another long molecule, in this case in the shape of a double spiral staircase. Each half of the spiral has a "backbone" attached to the outside of the "stairs," and each stair is one of four small molecules called bases: adenine

(A), thymine (T), guanine (G), or cytosine (C). In the double spiral, each A is across from a T and each C is across from a G. The sequence of bases in DNA is used as a "code" for the sequence of amino acids in a protein. A piece of DNA that is the code for one protein is called a gene. The code is that each sequence of three bases corresponds to one amino acid; the breaking of this code is one of the great achievements of modern biology.

How does this method of encoding get passed on? When a cell divides, which half gets the instruction book? They both do, because DNA is the code that DNA uses to reproduce itself. DNA is a double spiral with an A across from each T and a C across from each G. Each half of the spiral contains just the information needed to act as a template for the other half. Before a cell divides, the DNA double spiral unzips and each half is used as a template to re-create the other half. The result is two identical copies of the DNA double spiral, each of which goes to one of the cells produced by the division.

How do scientists know all this? Or to put it another way, what detection apparatus and methods have been used to find these things out? Though cells are small, they are much larger than atoms; so detecting an individual cell is much easier than detecting an individual atom. All that is needed to look inside a cell and see its workings is to get an image of a cell that is at least 100 times larger than the cell itself. A mirror reflects light and presents us with an image of the same size and shape as the object. But a curved mirror (like those used for makeup or shaving) or a lens focuses the light and can make an image that is larger than the object. The simplest microscopes (and here we return to one of our first examples in the introduction) consist of two lenses, the first to make a larger image of the object, and the second to act as a magnifying glass to give us a closer look at this image. Together, such an apparatus can easily yield the magnification necessary to look at cells.

But now suppose that we want to look in more detail at a single cell and see the different parts of it. Let's just make a more powerful lens with more magnification. This works, but only to a certain extent: light is made of waves, and waves have a certain wavelength. A wave cannot be used to resolve features smaller than its wavelength. This is a strange idea, in effect what it says is that light can be too big to see something by. Okay, so what we need to get a closer look at the parts of cells is light of a smaller wavelength, like ultraviolet rays or X-rays. Maybe we should build an X-ray microscope. Maybe, but then what would we use for a lens? An ordinary glass lens doesn't do so well at focusing X-rays. The answer to this dilemma lies in our TV sets (at least for those of us who haven't yet switched to plasma or LCD TVs). The heart of a TV picture tube is an "electron gun," a device that uses electric fields to accelerate electrons up to a certain speed and then direct them to a certain part of the screen. We have here a device for shooting and detecting electrons. We can build a better microscope by using electrons instead of light. Quantum mechanics tells us that electrons are waves also and that the more energetic the electron beam, the shorter the wavelength. So to get a beam of the appropriate wavelength, all we have to do is crank up the voltage on the electron gun. Furthermore, since electric fields can be used to direct the beam of electrons, a cleverly designed electric field can act like a lens and focus the electron beam just as the lenses in an ordinary microscope focus the light beam. This in a nutshell is the way that electron microscopes work and tells us they provide the resolution needed to study cells and the parts of cells.

While electron microscopes are vital tools in cell study, they were not responsible for discovering the structure of DNA. This was done using a nice example of indirectness in biology: a technique called X-ray crystallography. A crystal is a large structure of atoms packed together in a particular pattern. X-rays have wavelengths similar in size to the spaces between the atoms in crystals, and when a beam of X-rays bounces off of a crystal, it responds to the pattern of atoms in the crystal by producing a pattern of its own. Unfortunately, unlike the X-ray picture of a broken bone, in crystallography the pattern of the reflected X-ray beam looks nothing like the pattern of the atoms in the crystal. As an analogy, one can think of the crystal as a musical instrument and the reflected X-rays as the sound made by that instrument. X-ray crystallographers then can be thought of as listening to music and trying to figure out what sort of musical instrument made that music. This process cannot be done by detection alone: given only the X-ray patterns, it is not possible to calculate the crystal structure. However, given the crystal structure, it is known how to calculate the X-ray pattern. Thus X-ray crystallography needs a combination of detection and a theory that tells what a crystal structure should look like. This is like having a range of musical instruments that one knows the sounds of and then saying, "Aha, an oboe!"

In practice then, first the X-ray pattern is found, and based on that scientists guess a model for the crystal structure. This model can't be just any random guess because it must be consistent with what is known about the chemical composition of the crystal and with the chemical bonds that can form among those components. Then given a model, the crystallographers calculate the X-ray pattern that such a hypothetical crystal would form and compare it to the X-ray pattern of the actual crystal. If the pattern doesn't match, then it's back to the drawing board for the theorists until finally a successful model is found. This is the process by which the structure of DNA was found in 1953, where Rosalind Franklin measured the X-ray pattern of DNA and James Watson and Francis Crick produced the theoretical model.

The cells of our bodies are arranged in such an intricate and complicated fashion in interconnecting tissues and organs that it is natural to wonder how such an intricate arrangement came about. As with the question of the Sun, many theories were offered as to the origins of life. Most of these were, again as with the Sun, not truly concerned with the question so much as the use of the story to the people who heard it. Eventually scientific theories of life's origin were formed. One of them, Darwin's theory of evolution, has shown itself to work. This theory, which seems to be a triumph of pure thought, is also in a sense a product of a research program that has been going on for longer than written history. That research program is agriculture: the domestication and breeding of plants and animals. Before there was reading, writing, mathematics, or any other characteristic pursuit of civilization, there was applied biology. These people, of course, did not think of themselves as scientists. They were farmers and herders. But they knew by observation that within a given species of plants or animals, some had characteristics that were more useful (at least useful to the human beings who wanted to eat them). By selectively breeding the more useful individuals, they could (and did) generation by generation transform the appearance, size, strength, hardiness, and flavor of the plants and animals in their care.

Darwin called his theory natural selection to contrast with the sort of selection done throughout human history by farmers and herders; but one could just as well call this activity of farmers and herders artificial selection or controlled evolution. But why does controlled evolution work? Darwin reasoned that the information on what characteristics an animal has were inherited from that animal's parents, and he further hypothesized that that information is subjected to a small amount of random change. This hypothesis has been vindicated by modern studies of DNA, which put evolution on a firmer footing. It is DNA itself that is the inherited information, while the random changes, called mutations, are caused by damage to the DNA or errors in copying it. Darwin further reasoned that what artificial selection did is to make the information for the more desirable (to farmers and herders) characteristics of the animals ever more prevalent in the animal population; and that when by mutation an even more desirable characteristic (longer hair or more docility in sheep, larger size or a sweeter taste in apples, and so on) arose, that characteristic would also be artificially selected and become more prevalent in the population.

Given this way that the properties of animals and plants are inherited, what, Darwin asked, would happen to a wild animal population not subject to the artificial selection practiced by human beings? At first it might seem that the answer would be nothing. No artificial selection and therefore no change. But in order for an animal to pass on its genes, it must survive long enough to reproduce, and in the wild survival is by no means guaranteed. Thus even in the absence of artificial selection, it is still the case that certain characteristics are more likely than others to be passed on to the next generation. The characteristics likely to be passed on are whatever enables that particular animal or plant to survive in its particular environment: faster speed for an antelope, sharper claws for a lion, better camouflage for a chameleon, and so on. Even in the absence of the heavy hand of artificial selection, nature herself exerts her own natural selection and causes changes analogous to those produced by farmers and herders. Under the pressure of natural selection, the characteristics of animal and plant populations change. New species arise and old ones die out.

Notice that the language we use for this discussion contains the same dangerous kind of terminology as the discussion of "How does it know?" "Nature herself," we said, drawing on ancient personifications; "natural selection," we said, as if nature were a thing with volition. We use these terms and ideas because they are easier for us, but they carry the risks of distraction and confusion. We will talk more about this in the last chapter.

Mutation happens on a cellular level, producing individuals with new characteristics. These changes cannot in general spread sideways through a population; you cannot hand your genes to your neighbors (actually some cells can, but multicellular creatures like us do not). A new characteristic spreads forward in time, and perhaps becomes prevalent in a population yet to come. Look back at our light-cone illustration: it also does a good job of serving as a crude model for the possible spread of a successful characteristic. (This is only a crude model; population genetics produces much more accurate pictures.)

From this we see that evolutionary changes take many generations to fix themselves into a species. But Darwin lived in the time when the enormous age of Earth first became apparent, and further discoveries have placed his theory on an ever more solid footing. The fossil record shows the progression of life on Earth through vast swaths of time and the development of many different organisms. The study of DNA allows scientists to quantify how closely related different organisms are and to place all living organisms on one gigantic family tree.

Cells, DNA, and evolution seem very far from the cats and dogs and grass and trees of our daily life. Yet these three ingredients, found by the indirect means of detection and theory, form the foundation of modern biology.

As a shorthand scientific discussion of evolution, the above more or less suffices (particularly if one then dives into any of the hundreds of well-researched books on the topic; see the suggestions for further reading at the end of the book). Unfortunately, evolution is not only a scientific topic. It is also a religious, a political, and an aesthetic bone of contention (aesthetic because some people find the idea of evolution beautiful and others feel queasy at the idea of having nonhuman ancestors). Because evolution is treated in very unscientific ways in these other venues, we felt it necessary to add further discussion of it. We do not propose to debunk all the errors and misstatements made about evolution (that would take too long, and besides others have done it); rather we wanted to address one of the ways in which confusion is created: misuse of scientific terms.

Science, as with all fields of human endeavor, has words and phrases of specific meaning that either mean other things outside that field or simply have meanings that are not commonly understood. Sloppy use of such terms (either deliberately or accidentally) creates confusion (again either deliberate or accidental). We will have a broader discussion of this topic in the final chapter, for now having reached the perfect example on our return trip, we propose to stop and explore this rather disturbing vista. Don't worry—it won't delay our journey much, and we'll pick up some useful souvenirs along the way.

The first badly abused concept in evolution is that of "survival of the fittest." This phrase has been badly mangled to create the idea that there is an absolute fittest or best thing that creatures somehow are striving to become. But "survival of the fittest" really means that whatever fits the current situation is more likely to survive, even if in the long term the characteristics that led to present survival will eventually lead to the downfall of those creatures' descendants.

Let us take the case of sheep. We tend to think that intelligence and ability to defend oneself are survival characteristics. That is true unless one's species is being domesticated as food animals. In that case, these are the characteristics most likely to get an individual killed before it can breed. Wild sheep were much smarter and more aggressive than any sheep presently living on farms. The selection pressure (in this case shepherds) worked against those characteristics, culling out the dangerous sheep (a phrase not heard much these days) to make them more sheeplike (as we like to think of sheep). If humanity vanished tomorrow, sheep would have a great deal of trouble surviving because their fitness to survive in domestication is the same as unfitness to survive in the wild.

The other terms of confusion are usually coupled together: "random" and "mutation." Mutation is a change in genetic makeup so that a genetic characteristic appears in a child that does not exist in either of its parents. The most common form of mutation is an actual change in genetic code caused by an error in copying. If, for example, a piece of DNA contains the following sequence of nucleotides:

A A A A T T T T G A T C C T A A

and is accidentally miscopied as follows:

A T A A T T T T G A T C C T A A,

then the gene that is produced would code for the creation of a different protein than the original gene. This change could be irrelevant, minor, major, or fatal based on the function of that protein in the body. In a lot of cases, the mutation would simply prevent the new organism from developing at all, producing a stillbirth or the equivalent. A mutation only has a chance of entering the gene pool if the organism survives to viability. If that happens, the mutation undergoes the same of trial of life that every other organism does. The mutation might be useful or unimportant. It might begin as unimportant and only in later descendants become important. It might further mutate, producing unexpected effect. (The development of feathers into flight feathers is an excellent example of this gradual change from minor to major importance.)

Notice that the mutation undergoes two trials before it has any chance of propagation: first, whether or not the individual organism is viable, and second, whether or not it survives to reproduce. These two tests form a strong evolutionary filter against the random character of mutation.

It is time to take a closer look at randomness. While we touched on the concept before, particularly when looking at the quantum level of reality and the uncertainty principle, we need to look at it again on this level, where it is not so much a fundamental aspect of existence as it is a shorthand for an accumulation of events.

The classic example of a random act is the throwing of a die. Looked at as a physics problem, every toss of the die consists of an object put into motion in an environment. In theory the outcome of the throw is a solvable problem. Given the initial conditions, we could figure out exactly which face would come up when the die fell. In practice this problem is so messy and the initial conditions so hard to measure and so capable of small variation producing large effects that we cannot really determine the outcome.

Because we cannot predict the answer, the result of the die toss is deemed random. Mutation is like that, a series of events too minute and at the same time too numerous to calculate in any practical way. We cannot predict a mutation, nor can we easily determine its effect. We can look at the genetic code and see what has changed. With proper mapping, we can figure out how that changed gene will manifest in the particular organism. But what effect it will have on that creature's life we cannot easily predict.

Let us take a simple example of a mutation that has little apparent effect on survival chance. Suppose someone had a mutation in hair color that produced bright blue hair. Depending on the society the person grew up in, he or she might be shunned, or found attractive, or treated just like anyone else. In the first case, the chances of that gene being passed on are small; in the middle case they are high; in the last they don't affect it. That same mutation in a creature that unlike humans depends on camouflage for survival could have a more serious effect on its chances to reproduce, but the same three possibilities emerge. Depending on its environment, the new hair color could hinder camouflage, help it, or not matter.

Looked at from this perspective, the random quality of the survival of a mutation is a simple one. Does it hurt, help, or not matter? In the first case, survival is unlikely; in the middle, it is more likely; in the last, it does not contribute to the question. In other words, assuming an organism with the mutation is viable in the first place, and assuming the mutation does not actively prevent reproduction, its chances to propagate depend on how well the mutation works for the environment in which it lives.

The people who have trouble with random mutation as a mechanism for evolution usually say something like "It's impossible for complex life to emerge through a series of random events." But things aren't random in the way they seem to think. Every level of structure in the universe builds upon the levels below. At the base of life is chemical interaction, which works the way it does because of the quantum structure of atoms. Above the atomic level, life deals mostly with only a small set of elements: hydrogen, carbon, oxygen, and nitrogen. Other elements are used, but it is the chemical behavior of these four that determine what kinds of life are possible. Their chemical combinations, while multiple and varied, are nevertheless based on a small set of possible chemical bonds. The way these elements interact and can build up structures is impressive, but only a few possibilities lie at the root of that breadth of chemical variation.

There is a useful metaphor in the world of dice. If we start with a standard die, it can give us any number between 1 and 6. If we roll two dice, we can get between 2 and 12. Three gives us a range of 3 to 18, then four, 4 to 24, and so on. If we are rolling a million dice, we can get any number between 1,000,000 and 6,000,000. Yet that broad range of possible results is based on repetition of that basic 1-6 roll.

We can get a more extreme example if we take six ten-sided dice (numbered 0 to 9) and roll them without adding them to get the digits of a number. The first die gives us the hundred thousands digit, the second gives us the ten thousands, the third gives us the thousands, the fourth gives hundreds, the fifth gives tens, and the sixth gives the ones. This allows us to use only six simple random actions to get any number between 0 and 999,999. Again a small base of possibilities done over and over gives a plethora of possible results.

The crucial thing to understand about life in this is that it relies at each level of structure on a relatively narrow range of possible changes. It is the incremental change and aggregation of characteristics over time that lead to evolution. Single-celled animals did not mutate directly into human beings. They changed their genetic characteristics little by little, eventually producing cell colonies that rely on individual specialization. Then over billions of years, those colonies changed with survival pressure to produce descendants vastly more complex than their ancestors.

It is this combination of a large number of simple possibilities done repeatedly over time that turns the unlikely into the likely. In the above example of rolling six ten-sided dice, your chance of rolling all zeroes is one in a million. But if you made a billion rolls, you would expect a thousand rolls of all zeroes. If you look at the present state of complex life and say it's unlikely that randomness could lead to this state, you are right. But it is not unlikely that by rolling its many dice over billions of years, and at each roll keeping those combinations that work best, life could evolve to some form of complexity equal to ours. If you don't look at all the possibilities, you cannot see what is and is not unlikely.

MEDICINE

The science of life leads us naturally to that science that humans are at one time or another in their lives most concerned with: medicine. Medicine vies with evolution for the title of Oldest Practical Science. The history of medicine is also intimately tied into religion and magic, but this is a subject dealt with in other books that look more at the cultural basis of medicine than its scientific basis.

Medicine is an example of a narrow elaborate science that exists because of a special interest in one particular aspect of the universe. If we consider the sciences we have talked about—physics, chemistry, biology—they're very large fields. The divisions made in physics—quantum physics, atomic physics, mechanics, optics, electricity and magnetism, solid-state physics, planetary physics, astronomy, relativity, and cosmology—more or less correspond to different levels of structure in the universe or different questions that are of separate interest (with considerable overlap).

Yet in other sciences there are divisions caused by interest in particular subjects. Metallurgy as a branch of chemistry concerns itself with metals and has been of historical interest because of the use humans have made of metals. Farming as a branch of biology we have touched on, and again this is from human interest. Medicine comes about as a separate field from the larger study of animals for two reasons, first because we have a particular interest in keeping humans alive (or at least keeping one human alive), and second because medicine is the science that has the largest perceived universe.

It might seem as if astronomy and cosmology have the largest universe, and in terms of detection, you would be right. But every moment of every day each of us constantly experiences the perceived universe of medicine. We feel our hearts beat, we experience pain, we take in and exhale breath. We eat food and experience other digestion-related activities. Waking and sleeping, we are in the perceived universe of medicine. Medicine also has the most dramatic theoretical universe comprising as it does life and death, which on a cosmic scale are not much to sneeze at, but to us, they are very, very big.

Medicine is the end point of our journey, then, for two reasons. It brings us back to the perceived universe in a way no other science does, and it carries back bits of all the other sciences on our journey from astronomy, quantum physics, chemistry, and of course biology.

- For astronomy, consider the humble X-ray (not really humble, but picky about the light in which it is seen). The same band of electromagnetic radiation that reveals a black hole interacting with an accretion disk reveals broken bones.
- For quantum physics, the same kinds of accelerators used to test the theories of particle physics are used in radiation therapy for cancer.
- For chemistry, we are mostly C, H, N, and O. Their chemical reactions make us possible.
- For biology, we are living things in an ecosystem. We evolved and we are evolutionary pressures for the organisms around us.

Yet because it is the end of our journey, medicine finds itself facing concerns unlike those of any other science. The first such problem is ethical. It is simply immoral to perform a great many kinds of experiments that could tell us about human life. Many kinds of experiments on human beings would constitute torture. Medicine cannot and should not expand blithely into the detected universe. It may only experiment when it can act with primary regard for the health of the patient.

In most other branches of science, there need be no such concern. A nuclear physicist can tear apart atomic nuclei without any worries (unless too many are torn up in too small an area in too quick a time). An organic chemist can rend molecules asunder with happy abandon. A doctor cannot do such things to a patient. This means that medical knowledge and understanding expand very differently from that of any other science, largely growing in response to the pressures of whatever medical problems are prevalent. Medical knowledge evolves.

The second problem medicine has (and which it shares with other areas of biology) is that of isolation and repeatability, the hallmarks of the detected universe. While one electron is like another, one person need not be like another. Our bodies are complex interacting structures that use a multitude of different parts to perform even the most basic of functions. Consider a person who has trouble with digestion. Such a difficulty could arise in the mouth, stomach, liver, gallbladder, upper or lower intestines, in several hormonal glands, or in the nervous system; it could be an allergy, or it could be psychosomatic.

Separating factors and isolating causes are difficult in medicine. Two people showing the same symptoms might be suffering from two radically different ailments. Medical testing—knowing what to look for and what to test—is something of an art. Two equally trained doctors may be very different in their abilities to determine the cause of an ailment. Furthermore, two people of different genetic makeup may react differently to the same treatment. One gene can make the difference between recovery and relapse.

The third problem medicine has is one of long-term effect. A medical treatment may cause changes in the body that do not manifest for years or even generations. Human life spans are such that there is not sufficient time to make long-term studies for a new treatment, particularly if the treatment has been created to solve a major health problem here and now.

These three difficulties explain why medicine seems to suffer so many apparent reversals of good and bad treatments. What is seen as good in one decade may be seen as bad in the next because of long-term consequences. One generation's wonder drug has sometimes become the next generation's curse. In other cases, treatments have been overused or ignored because the treatment was only tested on one group of people. A common problem in medical studies has been insufficient breadth of genetic diversity in the tested subjects.

The difficulties that arise from the immorality of human testing have been equally bad, but in different directions. In some infamous cases, doctors have conducted human testing without the proper consent of the patients. This has been most commonly justified by the experimenters by treating the subjects as not really human. Studies of syphilis in African Americans and of the effects of radiation on developmentally disabled children came about by this form of dehumanization.

The second, equally dangerous, way medicine has handled this difficulty is by doctors being so enamored of theories that they do no proper testing, but jump in and use a treatment without con-

sideration of consequence. In all branches of scholarship, people can fall in love with theories and push them forward without regard for facts or testing. Medicine is the most dangerous field for this. Early twentieth-century treatments of psychiatric disorders included methods like lobotomies that fit somebody's theory, yet disregarded the facts that came out when the results of the "treatment" were examined.

We do not wish to sound as if we are down on medicine. We are not. Despite its difficulties, medicine provides the best argument for scientific education based on an understanding of theory and detection. Because this branch of science impacts our lives every day in ways that force us to make decisions and to ask informed questions, we cannot sensibly treat science as an abstract endeavor that has nothing to do with "real life." A wise person would want to know not just what a doctor assures them will work, but how they know it will work. If you are going to put your life or the life of a loved one in anyone's hands, you should need a little more than a good bedside manner.

As we said before, we have come to the end of our journey, having passed from the emptiness of dark energy back to the stars, to the planets, to our own planet, to the chemistry that makes life possible, to life itself, to our concern with our own lives, and arrived . . .

Where exactly are we?

We are back at the human scale, the place we left pretty early in this book. Here at the human scale, there is an instrument being used to detect and make observable certain characteristics of the universe. You are holding it in your hands (unless you're reading this on a screen). It's this book. Books on science can be thought of as instruments of the detected universe. They are indirect tools for gathering information about the universe.

Books and articles on science are the most commonly used tools for this purpose. Scientists use them all the time. Books and articles allow them to pass around and pass on knowledge and the results of experiments. They ensure that each scientist does not need to reinvent the wheel or redo the experiments that determined the diameter of Earth, the length of the AU, the speed of light, the curvature of the universe, and so on. They also ensure that should a scientist want to redo any of these, he or she will know how.

But how can one be sure that such an instrument works as it claims to?

In short, "How does the book know that?"

CHAPTER 14

Science as It Is Written

Neither of us has ever written a science book before; but as a scientist and a writer, we collectively know something about science and something about writing. It might seem that putting these two together would be an easy task, but there is a fundamental conflict between science and writing that is rarely acknowledged. Writers strive to create works that flow easily, so that readers can simply sit down and take in the images and ideas the writer has sought to craft. Even if the ideas are strange and the images outlandish, a goal of writing is easy absorption on the reader's part. In science fiction one tool of this is that the writer should encourage in the reader what is called suspension of disbelief. That is, even if what one is writing is ludicrous by real-world standards, the writing should create a world that seems so natural to the reader's mind that they will take in violation after violation of the laws of nature.

Science, on the other hand, demands the opposite. In learning science, one must have suspension of belief. One must strive for an extreme skepticism and an insistence on the answer to the question "How do you know that?"

On a smaller scale, writing works by flowing one idea after another, usually sentence after sentence that builds up an image in the mind. Smooth flow is considered a characteristic of good writing, and editors scream about sentences that break up that flow. Science, on the other hand, has a more staccato rhythm, a challenge-and-response, question-and-answer rhythm. That might sound like it flows, but there is an inaudible beat in the back-and-forth of science: Challenge, pause and think, response. Question, formulate experiment, create experiment, perform experiment, answer. Sometimes the inaudible beats last lifetimes.

Up until this chapter, we have been striving for one of the fundamental goals of writing: smooth communication of ideas. In so doing, we have leaned far on the side of writing because we relied on what writing does well in order to communicate some pretty difficult science. We have used description, imagery, metaphor, humor, asides, biography, history, and even a little conflict for our ends. We have also relied on the fact that of all the arts writing is best at getting inside people's heads. Only in writing can one simply stick a thought down and show it to the reader. (Try doing that in a movie, a painting, or a sculpture. It's really difficult.) We exploited this advantage of writing to discuss not just the science but the ways that scientists think.

Here and now we are going to stop (for a time) deferring to the advantages of writing, because we need to outline its disadvantages. We need to do so because as we said the ways of writing are in conflict with the ways of science, and that conflict is sometimes a great drawback for science. In that conflict, science is most often the loser. The suspension of disbelief is much easier for most people than the suspension of belief. This is just fine if one is reading fiction, but not if one is supposed to be reading fact.

Indeed, we have presented as fact a model of scientific thought that most scientists will acknowledge as true, but that, as far as we know, has never been offered in quite this way before. The three-tiered universe is our own invention, a written image of a shared way of working and thinking. But we have not offered, nor can we offer, proof that this is the way each and every scientist thinks about his or her work. It isn't. It is an image of the generally shared manner of careful thought that scientists on the whole use. But the image itself and the terms of the three-tiered universe are new.

Furthermore, the borderland between fact and fiction is not a sharp line but a blurry area. The old-fashioned historical novel and the new fashions in docudrama and infotainment have routinely sacrificed accuracy of portrayal in favor of dramatic or humorous or fashionable images of people and ideas. Science writing is not safe from this blurring. A great deal of it sacrifices scientific precision for the sake of a writer seeking to make a story "more interesting." This again shows a basic conflict. To a scientist, there is inherent interest in trying to unravel the way the world works, not just the whole overarching cosmological paradigm, but a little bit of it. An entomologist may spend a full lifetime discerning the ways of a single species of insect and be praised for the care and thoroughness of such work. A theoretical physicist may devote a lifetime to working out the consequences of a theory so that a test can be made that disproves that theory. To a scientist that is a lifetime well spent. But in standard conventions of storytelling, these well-spent lives would be deemed tragedies.

This may sound like a minor conflict of aesthetics (unless you're the person whose life is being belittled), but there are more serious problems of science writing that confront anyone who wishes to learn about science without actually having to be a scientist. Those problems arise in books and articles about science—in short, in works like this. The problem fundamentally lies in the blurring of science for the sake of entertainment. This produces indifference to the realities of the real world, because it is through science, as we said at the beginning, that we get at so many of the facts. But what happens if science as presented becomes nothing more than a form of entertainment. Well, to quote Cole Porter:

Have you heard? It's in the stars.
Next July we collide with Mars!
Well, did you ever?
What a swell party this is!

Of course, these lyrics are a complete caricature: two people so brainless that they have no idea that the "news" that they are passing back and forth would, if true, mean the imminent demise of themselves and everyone that they know. But this caricature pinpoints a crucial shortcoming in many science news stories: These stories, when we pass them along, don't equip us to say much more than "Scientists say . . ." or "Scientists have found . . ." In short, what's missing is that the stories don't provide an adequate answer to the question "How do they know that?"

"How do they know that?" is both the heart of science and the ultimate trip-up question, a conversation killer, a ripper-up of plots, and a reason for editors to yell at writers. Nothing would more destroy a story than to have to explain exactly where everything came from and why things are the way they are. In other words, we are crazy to have written a book centered around something that messes up books. We're crazier than that. We think there needs to be more of this question. Not in fiction, where explanation is only useful when it aids flow, but in nonfiction, particularly scientific nonfiction.

This may sound like the pet peeve of a few disgruntled scientists. What difference does it make if most people see science as just another light show on their televisions or just another story in their magazines? It makes a great deal of difference because, as we showed in the previous chapter, science is not divorced from everyday life, and the standards of entertainment are poor methods of selection for serious things. You do not, for example, want your doctor to pick treatments for your illness based on which ones have the most news coverage or the best special effects.

Of course, in a short article such as might appear in a newspaper or on a television news show, space is severely limited, but we think that at least some of that space could be put to better use in giving more information about the techniques of detection. This is not only because the detected universe is an important part of science, but also because detection techniques are common to many different experiments, many different fields of science, and many aspects of everyday life.

For example, we have previously discussed the Doppler effect and how it is used to measure speed. Coupled with Kepler's laws, it allows astronomers to find the masses of stars in binary systems and to find the presence and some properties of extrasolar planets. Through the Hubble law, the Doppler effect allows astronomers to find the enormous distances to quasars and supernovae. On a more day-to-day level, the Doppler effect is also how police measure the speed of cars and how meteorologists measure the speed of clouds. The Doppler effect, if well reported each time it is part of the detection technique of a science news story, could knit together for readers several diverse parts of science and also connect science to everyday life. No longer would each story be an isolated fact, and in passing along the story, the reader would no longer be confined to "Scientists say that . . ." but could instead say, "Here's how it works . . ." And if there is something strange in the report, a reader could say, "Huh, I did not know the Doppler effect could be used for that." And then, "Maybe I should find out more."

The presentation of science as a set of isolated facts leaves out the beautiful and practical connections between different parts of science that are the hallmark of the detected universe. Isolation cripples understanding and use of detection. It also leads to a very limited and distorted picture of the theoretical universe. Many of the distortions have to do with the ambiguity in the word "theory." Consider the distinction between fact and opinion: Facts are true, while opinions are things that some people think are true, which might or might not be. It is tempting to fit science to this mold by thinking of all experiments as fact and all theories as opinion. But scientific theories are tested by experiment. The ones that fail the test are discarded, while the ones that pass have achieved a stronger status than they had before, more "fact-like" and less "opinion-like." It is probably best to think of the status of a theory as being somewhere on a continuum with "wild speculation" on one end and "fact" on the other. Very well-tested theories, like the atomic theory that matter is made of atoms, should be thought of as fact.

Unfortunately, none of this subtlety is captured in the word "theory," nor can this problem be readily solved by a change of terms. Since the status of a theory slowly changes as the experimental evidence mounts up either in its favor or against it, there is no particular place where the status of the theory dramatically changes from opinion to fact.

The situation is different in mathematics. Here there is an incontrovertible demonstration of mathematical fact, the mathematical proof. A mathematical idea for which there is no proof is called a conjecture, which is the technical mathematical term for mere opinion. Once there is a proof, the mathematical idea is called a theorem, the technical mathematical term for a fact. No other field than mathematics has this pure and absolute a black-and-white distinction. Everything else uses the continuum.

This ambiguity in the term "theory" creates potential for storytellers to pull out their tools and create articles that, while not necessarily nonfactual, create false impressions for their readers. Theory in particular leads to three types of pitfalls that can occur in science news stories: (1) the exaggeration of a controversy, (2) the overdramatization of an experiment, and (3) the overselling of a speculation.

An example of pitfall 1 is found in discussions of the age of the universe. Before the discovery of dark energy, the estimated age of the universe based on big bang cosmology seemed to be younger than the estimated ages of some stars in some globular clusters. The experimental results were not good enough to say that there was a definite disagreement, but the theories of cosmology and stellar structure seemed on a collision course. One or the other (or both) of these theories would require some reworking. This was an interesting issue and was reported in various newspapers. But a number of stories managed to give the impression that the whole big bang theory was in trouble and that this controversy might bring it crashing down.

There was less reporting explaining that the big bang theory is on solid ground and supported by many detailed astronomical observations (as we discussed with the observation of galactic motion and the CMB), and that therefore any resolution of the issue would have to involve tinkering with the big bang theory rather than discarding it.

A nice illustration of pitfall 2, overdramatization, comes in stories about the observation of black holes. As a concept in astrophysical theory, black holes were initially regarded as quite exotic, so it took awhile for them to be accepted as the standard explanation for such observational phenomena as QSOs and active galactic nuclei. There is no sharp line that a theory crosses to become a fact, no specific test that ends all debate. Nonetheless, stories on black hole observations are often written to imply that this observation is the one that finally nails things down and shows that black holes really exist. In fact, it is the aggregation of detection that leads to acceptance. No one result creates acceptance in science, since any experiment can have hidden flaws. You need a mass of data from multiple sources, not just one hopeful experiment.

A helpful illustration of pitfall 3, the overselling of a speculation, comes from a group of reports (still going on at the time this was written) about the possibility of producing black holes in particle accelerators. It certainly sounds impressive to say that, according to scientists, a particle accelerator in a big underground tunnel in Switzerland may start making tiny black holes. However, the content of the science in this case is that *if* a particular completely speculative and totally untested particle physics model turns out to be true, and *if* the completely unmeasured parameters of this model turn out to have certain values, then black holes *might* be produced in the world's largest particle accelerator. And those are two gigantic *ifs*. They are not as gigantic as *If there is a monster asleep under an island and if an atomic test wakes it up, then the monster will go on a rampage destroying Tokyo, whose citizens for no clear reason are speaking in badly dubbed English, unless you get the director's cut.* But they're still pretty darn big.

We don't want to be seen as picking on the physicists working on these theories. They are reasonable scientists following an interesting theoretical line of thought. But the reports, by not explaining just how speculative that line of thought is, make the whole thing seem much more dramatic than it really is.

The problems outlined above sound fairly minor, since the above news stories concern scientists honestly following their professions and trying to work out subtle problems in their fields; though in the stories about them, there is a push toward "better storytelling" that creates dubious impressions. However, this type of science storytelling does not help the public become scientifically literate. So what's the big deal about not having a scientifically literate populace? What's at stake here? A great deal, as it turns out. In today's democracies, citizens (mostly through their elected representatives) participate in decisions on a wide range of policy issues with a science component, from global warming to stem cell research.

But more than that, we are in daily life, through political tracts and in advertising, exposed to "stories" of a more disturbing sort, that claim to be science but are not. Scientific literacy and care are needed on the reader's part because such stories are being used not just for drama but to deliberately mislead, whether the purpose is to deny evolution or global warming, or to sell the latest quack "nutritional supplement."

To demonstrate the seriousness of this matter, we have to delve at least a bit into the unpleasant topic of pseudoscience.

There is this important principle of writing: The truth or falsehood of what one is writing has no effect on a writer's ability to make it believable. This is because believability is a subjective judgment on the reader's part, based on what the reader is comfortable with accepting and what he or she has come to accept as a truthful presentation.

Writing is not alone in this. Any art can make something unreal seem real. As this is being written, there is a beautiful image of a winged human figure on the computer screen. That figure has no basis in reality, but it seems real. Painters, sculptors, puppeteers, and animators strive to create objects and creatures that will be accepted as real and right-looking by their viewers. This is one of the things that art does. It creates a real-seeming unreality. There's nothing wrong with this so long as the real and unreal are firmly labeled as such. But long ago people discovered that one could use these arts to muddy the waters between the two. So we have propagandistic images and writings. We have doctored photographs, edited video and audio, and so on.

Within itself, science strives to make sure it does not fall prey to such manipulations by using appropriate ways and methods to check the reality of what it is doing. There have been times and cases where it has fallen down on this job, but it has corrected these errors in time and fixed things.

Outside of science, science is seen as this strange world with weird people in it, fodder for stories and for propaganda. It is important to remember that the most common target for propaganda is those weird people over there. Aren't they disturbing/fascinating? You want to do/don't want to do what they do.

Despite our previous complaints about articles, we have reached an area where science and journalism are, in principle,

natural allies. Both seek to reveal the world as it is, both strive for presentation of the truth, and both decry the falsification of evidence. Journalism and science are both seen to punish their own only in cases of falsification. But there's falsification and then there's "telling all sides of the story." The latter sounds very good to journalists. If one person says one thing and another says the opposite, that makes for a "good story," and if the truth is not obvious, if the people telling the story lack the means to tell one side from the other, then many journalists will default to telling both sides, or if one side is commonly accepted, then the existence of another side of the story makes for a good story.

In order to be able to tell whether one should or should not treat two sides of a scientific dispute as equal, it is necessary to have a means of discernment between them. We have proposed in this book the scientific standard of "How do you know that?" Following this standard involves longer, more painstaking articles on the part of the writers and a greater interest in the nuts and bolts of things on the part of the readers. This may sound like an unnecessary burden. But consider what happens if this standard is not followed. If the standards of writing are used, drama and appearance trump technical reality and careful analysis, and appearance overwhelms detection. Down this road lies the path to legitimizing junk science and pseudoscience, as we shall see.

Science is a method for interacting with the world in such a way as to discern the nature of the world. Scientists do this using the three-tiered universe with the flow of observation, theory, and experiment that we laid out. But how does science look to most other people? What do scientists *appear* to be doing while the universe is buzzing in their heads? This may seem an odd question, but in the same way that an actor can take on the role of a completely unlike person, a nonscience can by imitating the appearance of science pretend to be a science without having any science in it. Why do this? Because science has believability and the appearance of science can be used to convince people of many things that are wholly untrue.

Let us look at the trappings, the costumes, props, and scenery of science, in order to understand how such a masquerade can be perpetuated. The simplest of these props is the word "scientist." Many reports of new ideas and studies begin with some variation of the phrase "Scientists have determined . . ." followed by an encapsulation of the conclusion of a supposedly scientific study with little reporting of content or method or even who these scientists are. The scientists may be botanists who have a new cosmological theory that has not even been looked at by cosmologists. There is no broadly accepted definition of what makes a person a scientist. A degree in science is useful, but there are also amateur scientists who have in the past done much important work. So the word "scientist" can be slapped on to anyone.

Related to the word "scientist" is the title "doctor." Doctorates can be given in nearly any field. A person with such a title may be a world-renowned expert in Bach's music, but that doesn't mean that his new theory of the fundamental structure of the universe should be given any more weight than that of some un-degreed person ranting in the streets. Doctorates represent expertise in one field. The question must always be asked, "Which field?"

Other common trappings in the appearance of science are the names of institutions, pictures of equipment, and the old reliable prop for movie and TV scientists, the lab coat. White lab coats have been used to make people look like scientists for nearly a century. Nowadays, of course, most real scientists—except medical doctors—don't wear the things and are more commonly found in street clothes. (We will not discuss the general level of sartorial skills among scientists.)

The most serious prop and the hardest for the layperson to deal with is the use of mathematics, particularly statistics. Most people's mathematical experience boils down to artificial problems assigned to them in schools that have simple right and wrong answers. This gives even the most math-phobic person a naive faith in numbers that the experienced mathematician or scientist does not share.

The ugly truth is that it is always possible by dishonest manipulation of numbers to make the results of a study appear to be whatever someone wants them to be. Several books have been written about this kind of numerical trickery, so we will not belabor the point, but we will offer one example: Suppose one wanted to know what houses in a certain area were selling for, on the average. Suppose further that there were five houses selling for $200,000, five selling for $300,000, and one mansion in the area that sold for $10,000,000. What would the average price be? That would depend on what one means by the word "average." Statisticians use several different kinds of averages in their work. The most common is called the mean and is calculated by adding up all the numbers and dividing by however many numbers you have (in this case eleven numbers for the eleven houses). That gives

$$[(5 \times 200{,}000) + (5 \times 300{,}000) + 10{,}000{,}000]/11 = \$1{,}136{,}364$$

Thus a person calculating average price this way could say, "The average house here sells for more than a million dollars." This, while apparently true, is highly misleading.

A better indicator of price is an average called the median. Housing prices are, in fact, usually listed using this kind of average. For the median, you list all the prices in order from lowest to highest and pick the one in the middle of the list. In this case this gives us an average of $300,000, which is more truly representative of the prices in this area.

Most scientific studies, particularly those of complex subjects, contain so many different numbers that this kind of manipulation can easily be used in such a way that it is hard to discern what is really going on, particularly if the report does not tell the methodology used to reach the conclusion. If you write a paper by starting with an explanation of the problem, explain your answer, and pad the paper out with terminology and theory without ever dealing with detection and calculation, you can create the illusion of a scientific report. You won't fool the experts, but you can fool a lot of other people.

This leads us to the unpleasant topic of junk science. Junk science is science done to establish a preconceived notion—not to test the notion, which is what proper science tries to do, but to establish it regardless of whether or not it would hold up to real testing. Junk science reports are in effect written backward. The conclusion is decided and then the experiment that will establish it is done. Junk science is ugly and pernicious and has three causes: attachment to theory, biased study selection, and plain old lack of ethics.

Attachment to theory comes about when someone falls in love with an idea and will not let it go regardless of the evidence. Such a person will do anything to show that this idea is correct. This is the academic form of a dysfunctional relationship. The academic gives everything to the theory; the theory gives nothing back. There have been many such tragic love affairs in the history of science. Einstein's objections to quantum mechanics because of its random character is one of the better-known examples. We also noted Einstein's self-confessed greatest blunder. This is particularly ironic, considering how many people refused to listen to Einstein because of their own attachments to Newton's theories.

Biased study selection is related to attachment to theory but more commonly has an institutional cause. Rather than deciding on the outcome and creating the experiment beforehand, the biased selector simply picks those among other people's results that fit his or her theory. Biased study selection can also happen with attempts to discredit undesired science. The corporate smear campaign against early environmentalist Rachel Carson during the 1960s falls into this category.

Further away from real science than junk science is pseudoscience. The hallmark of pseudoscience is the after-failure justification. When a study or experiment fails to produce the desired results, a pseudoscientist comes up with an explanation that says why things didn't work this time. This explanation will not be determined by careful study of the experiment or the creation of a new experiment. Rather, the explanation will be used as a means to justify throwing out the offending result. In other words, pseudoscience is characterized by not paying any attention at all to whatever goes against one's ideas, dismissing things out of hand. This is a more extreme form of attachment to theory than that found in junk science, in the same way that a gunfight is an extreme form of roughhousing.

An example of pseudoscience: The failure of Soviet agriculture in the 1940s and 1950s can be attributed to the failed theories of a biologist named Trofim Lysenko, who thought that acquired characteristics—not just genetically determined ones—could be inherited. This theory of evolution was vigorously supported by Stalin because it agreed with communist ideals, although not with actual observation. It led to disastrous crop failures and famine in places that had once served as breadbaskets for vast regions. This is one of the cases where political and military force has been used to back up attachment. Many biologists who spoke out against Lysenko were exiled or sent to labor camps. Only after many years of suffering was the theory removed from prominence; the consequences in the universe of fact, however, have not gone away.

There are many books dealing with pseudoscience and a fair number of people and institutions who devote time and effort to debunk various pseudosciences (we mention a few in the further reading suggestions), so we need not dwell on it here. We would rather instead devote our remaining efforts to discussing how one can gain information on real science, despite the superficial way that science news stories are often presented.

BEYOND COMPLAINING

What can be done about the problems with newspaper science stories? In the previous section, we outlined a solution from the side of the writers: We suggest that stories say more about the detected universe and give more thorough answers to the question "How do they know that?" However, we have no delusions of grandeur, no notion that many reporters will actually deign to take our advice. After all, they have learned and practiced their craft for many years. Why should they listen to a scientist and a writer with the colossal nerve to tell them how to do their job?

We are therefore going after the other side of this relationship: readers. We are making a suggestion to people who read science books and articles. We suggest suspension of belief and strong wondering about "How do they know that?" If they don't tell me how they know that, why should I believe them?

That may sound like nothing more than an internal protest against what is presented to one, but it actually points toward what an interested reader can do beyond complaining. The first step of this process after the initial how question is to recall that though these days we use the word "media" simply as a synonym for "newspapers, radio, and TV," it has a more general meaning. Media (singular: medium) is simply another way of saying middlemen or go-betweens. In this case, the newspaper science stories are the middlemen between scientists and the readers of science. This naturally leads to the question: If we're not happy with the job done by newspapers as go-betweens, then what other media are available? The possibilities we will consider are professional scientific journals, science magazines, books, and the Internet.

Scientists tend to publish their work in professional scientific journals. In fact, it is a cliché that in academia the choices are "publish or perish." However, the reports that scientists share among themselves are nothing like the reports of science that reach the general public. This is not because of a desire to keep secrets, but rather because people who share a common profession can talk to each other in a shorthand language that may be a quick and an efficient form of communication, but is often confusing for those outside the field who do not know the terminology. Using this quicker form of communication, scientists can send e-mails a few lines long that contain a huge amount of explanation. When a new theory has been propounded or a new experiment done or a new thing discovered, scientists will write up explanatory papers that are highly technical and very specific in content. Unfortunately, these papers are totally incomprehensible even to scientists in slightly different fields. Clearly professional scientific journals are not a suitable medium for communication between scientists and readers of science.

Science magazines are meant for the scientifically educated layperson. They are written at about the depth of the kind of book you are now reading and presume a certain level of sophistication on the part of their readers. Their writers are often scientists who think their work or discoveries might be of general interest. The articles usually include some details of experimental techniques and theoretical underpinnings. Regrettably, the writing in these articles is sometimes rather dry because scientists are usually not trained writers. Despite this stylistic drawback, it is usually in these magazines that one can find news of discoveries that answers the question "How do they know that?" These can be the best sources for readers of science and for scientists reading far outside their fields.

As for books (more specifically science popularization books), clearly we are in favor of this medium since it's the one we've chosen. It is encouraging that many scientists and writers choose to write these books. We hope that some of them will find our categories of the perceived, detected, and theoretical universes useful. And we expect that even more will find the question "How do they know that?" and its answers a helpful method for presenting science. In hindsight, perhaps the most useful thing that we can say about this means of communication is that it's a lot harder than it looks. (Don't get us started, please.)

Unfortunately, none of these media address the following situation that we often find ourselves in: We read a newspaper science article and find the subject interesting but are disappointed by the small amount of information provided and in particular by the lack of an adequate answer to the question "How do they know that?" Where do we go for more information on that specific subject? There may not be any science popularization books on the subject, and even if there are, we may not want to read a whole book just to get the answer to this one subject. There may not be any articles in science magazines on this subject, and even if there are, how do we find them?

There's the Internet. There is a lot of information on the Internet, yet it is as wide open and lawless as the Wild West of American legend. Can we find what we are looking for on the Internet? And if so, where do we look? We want the information to be reliable, but we also want to be able to find it easily. The good news is that there is reliable information on the Internet and that information on the Internet can easily be found. The bad news is that there is something of a trade-off between these two criteria. Let us examine three sources of Internet science information: outreach websites, wikis, and search engines.

"Outreach" is the term used by scientists for their efforts to explain their work to the general public. Though websites are not the only form of outreach (public lectures are another), they are the most easily accessible. Individual scientists may post outreach materials on their own websites. But large groups of scientists, in particular science institutes or large experimental facilities, are more likely to have an organized outreach activity, often with one or more staff whose job it is to coordinate and maintain the outreach website. Such websites are the most reliable of Internet science information. However, if one is searching for a particular piece of information, it is not obvious which outreach website is likely to have it nor how to find that particular website. Furthermore, as with the image of a scientist, it is perfectly possible to make a website that looks like an outreach site but that pushes an agenda, a pseudoscience, or any of the other science-spoofing things we mentioned previously. It is vital if one has found such a site to find out who and what is responsible for it. This is particularly true in the case of controversial subjects.

"Wiki" is a term for a website that allows its visitors to add to, edit, and change its content. In principle, this is an extremely efficient and rapid way to amass information, especially in new and rapidly changing fields. However, also in principle, this is a system in which it is hard to guarantee the quality and accuracy of the information so amassed. Ultimately the quality of a wiki depends on the expertise and good intentions of its contributors, and on the quality control exerted by the administrators of the website. We have not made any systematic study of the subject, but our own experience of wikis as a source of science information has generally been very positive. For many wikis, it is possible to not only read articles but also the discussions of contributors so that one can get a better idea of why certain ideas appear in the articles and certain others do not. The discussions can also be entertaining.

Search engines simply take a term entered by the user and search for websites in which that term occurs. This is the easiest way to search for information. But by their nature, search engines cannot vouch for the accuracy and quality of the content of all the websites that they find. Checking the who and what of a site will help greatly with finding out whether or not it has reliable information. In any case, among the other things that a search engine finds will often be the relevant science outreach websites.

Whatever medium is used, it is important that scientists and readers of science not feel constrained by the methods and conventions of newspaper science reporting. Science as it is really done is far grander and more beautiful than science as it is often seen. The presentation of science as a bunch of disconnected instances of "Scientists say that . . ." cannot hope to capture the intricate and interconnected web of knowledge that consists of the perceived, detected, and theoretical universes, and that is elucidated in the answers to the question "How do they know that?"

Looking at one new idea alone is like hearing a single note, a bright sound in the mind. But listening to that note in context with the other notes of knowledge, one will hear more than any of them can sing on their own. The universe is harmonious. At every level of structure, the things of that level—fields, particles, atoms, molecules, cells, life-forms, stars, galaxies—interact to create new things. This applies to the mind as well. One idea can be bright and amusing, a good story, a "Well, did you ever?"

But many ideas working together in harmony, many little bits of "what" tied together with pathways of "how," can grow whole new levels of understanding that last far longer than a single interchange or a single article. The oldest bits of science in this book are older than human history. The newest are being worked on right now. But they are all threaded together and make a single thing that itself seeks out to touch and harmonize with other ideas. Science reaches outward to the world of fact through its three universes and inward to the mind by offering its three universes—its answers, its questions, and, most vital of all, its ways of going from question to answer to the next question. That's how they know that.

GLOSSARY

absorption lines. Dark lines in the spectrum of a diffuse material that come from particular wavelengths of light being absorbed by the atoms of that material.

accretion disk. A donut-shaped cloud of gas in orbit around a white dwarf, neutron star, or black hole.

active galactic nucleus. Also called AGN. The central region of a galaxy that is emitting large amounts of electromagnetic radiation due to matter from an accretion disk falling into a supermassive black hole.

antiparticle. For each type of particle (for example, the electron), there is another type of particle with the same mass and opposite electric charge (in this case, the positron) called its antiparticle. The antiparticle of the proton is called the antiproton. The antiparticle of the neutrino is called the antineutrino. The photon is its own antiparticle.

arc minute. 1/60th of a degree.

arc second. 1/60th of an arc minute (1/3,600th of a degree).

astronomical unit (AU). Average distance from Earth to the Sun.

big bang. The explosion at the beginning of the universe.

black hole. An object that has undergone complete gravitational collapse to form a region where gravity is so strong that even light cannot escape.

blueshift. The decrease of the wavelength of light from an object that is moving toward us.

Chandrasekhar limit. The maximum mass of a white dwarf, about 1.4 solar masses.

cosmic microwave background (CMB). The light left over from the big bang, now chilled to a temperature of about 3 kelvin by the expansion of the universe.

dark energy. A substance that is gravitationally repulsive and is responsible for the acceleration of the expansion of the universe.

dark matter. Matter that is not seen and is only detected indirectly by its gravitational effects on the motion of stars in galaxies and of galaxies in clusters.

degeneracy pressure. A quantum mechanical pressure that electrons (and neutrons) exert when many of them are confined to a small space. Degeneracy pressure is due to the uncertainty principle and the Pauli exclusion principle.

Doppler effect. A change in the wavelength of a wave due to the motion of the source (or of the observer). The Doppler effect is responsible for the phenomena of redshift and blueshift.

electromagnetic force. The force that electric and magnetic fields exert on particles with electric charge. (Electrons and protons have electric charge. Neutrons and neutrinos do not.) Since charged particles make electric and magnetic fields, one can also think of the electromagnetic force as the force that charged particles exert on each other.

electron. One of the subatomic particles that make up the atom. An atom consists of a nucleus and one or more electrons with the negatively charged electrons in orbit around the positively charged nucleus.

event horizon. The event horizon of a black hole is the boundary of the region from which light cannot escape. Anything within the event horizon is in the black hole interior and cannot get out.

Fraunhofer lines. Absorption lines in the spectrum of the Sun.

general relativity. Einstein's theory of space, time, and gravity. In general relativity, gravity is due to the curvature of spacetime.

gravitational force. A force that all masses exert on each other. This force is well described by Newton's law of gravitation, except where gravity is so strong that one must use general relativity instead.

Hawking effect. The emission of thermal radiation by a black hole.

H-R diagram. A graph of stars giving their temperature and luminosity.

Hubble's law. The law that states that galaxies move away from us with speeds proportional to the distance from us. The constant of proportionality in this law is called the Hubble parameter. Hubble's law is a consequence of the expansion of the universe.

Kelvin-Helmholtz mechanism. The heating of a gas that is contracting under the influence of its own gravity.

Kerr's formula. Usually called the Kerr metric. The mathematical description of a spinning black hole.

main sequence. That part of the H-R diagram where stars are for the main part of their lifetime: the time when they are fusing hydrogen to make helium.

neutrino. A subatomic particle that has zero electric charge and a very small mass and that interacts very weakly.

neutron. A subatomic particle that is one of the constituents of the atomic nucleus. A nucleus is made of protons and neutrons. The neutron has zero electric charge.

neutron degeneracy pressure. *See* degeneracy pressure.

parallax. The apparent change in position of an object under a change in the point from which that object is observed.

parsec. From parallax-second. The distance of an object whose parallax, when seen from opposite sides of Earth's orbit, is 1 arc second.

Pauli exclusion principle. The statement that a given quantum state can contain no more than one electron. Also applies to neutrons and several other types of particles.

photon. A particle of light.

proton. A subatomic particle that is one of the constituents of the atomic nucleus. A nucleus is made of protons and neutrons. The proton has positive electric charge.

quasi-stellar object. Also called QSO. An extremely powerful (and usually extremely distant) active galactic nucleus.

red giant. A star that is generally much larger and colder than main sequence stars. Red giants are fusing helium rather than hydrogen.

redshift. The increase of the wavelength of light from an object that is moving away from us.

Schwarzschild radius. For an object that is not a black hole, the Schwarzschild radius is the size it must shrink to in order to become a black hole. For a black hole, the Schwarzschild radius is the distance at which gravity is so strong that light cannot escape.

spacetime. The combination of space and time. Spacetime is the collection of all events, where an event is a single point of space at a single moment of time.

stellar mass loss. The ejection of matter from the outer layers of a star due to the intense radiation generated by the star.

strong force. The force that holds protons and neutrons together to make the nuclei of atoms. This force is much stronger than the electromagnetic force.

supernova. The explosion of a star. A type Ia supernova occurs when a white dwarf undergoes a sudden thermonuclear explosion as accretion makes it approach the Chandrasekhar limit. A type II supernova occurs when the iron core of a massive star reaches the Chandrasekhar limit and collapses, turning the iron into neutrons and causing an enormous outpouring of neutrinos.

uncertainty principle. The statement that the accuracy with which position is measured limits the accuracy with which speed can be measured. One consequence of the uncertainty principle is that confining particles to a small region makes them move rapidly.

Unruh effect. The absorption of thermal radiation from the vacuum by an accelerated observer.

weak force. The force responsible for nuclear beta decay and in general for any nuclear reactions involving neutrinos. This force is much weaker than the electromagnetic force.

white dwarf. A star, made of carbon and oxygen, that is not undergoing nuclear fusion and is held up by electron degeneracy pressure.

SUGGESTIONS FOR FURTHER READING

Our favorite popularization of black holes is *Black Holes and Time Warps*, by Kip Thorne (W. W. Norton, 1994).

The foundations of general relativity are well covered in *Space, Time, and Gravity*, by Robert Wald (University of Chicago Press, 1977).

For an introduction to the big bang, we recommend *The First Three Minutes*, by Steven Weinberg (Basic Books, 1993).

Particle physics, in particular the standard model, is covered nicely in *The Theory of Almost Everything*, by Robert Oerter (Pi Press, 2006).

The discovery of the structure of DNA is recounted in many books of which our favorite is *The Double Helix*, by James Watson (Atheneum, 1968).

A beautiful treatment of evolution can be found in almost any book by Richard Dawkins, of which our favorite is *The Selfish Gene* (Oxford University Press, 1976).

Pseudoscience is treated in *Voodoo Science*, by Robert Park (Oxford University Press, 2000); *Fads and Fallacies in the Name of Science*, by Martin Gardner (Dover Publications, 1957); *Science Good, Bad, and Bogus*, by Martin Gardner (Prometheus Books, 1990); and "106 Science Claims and a Truckful of Baloney," by William Weed (*Popular Science*, May 2004).